Geological T

D1237734

EVENTS

MILLIONS OF YEARS BEFORE PRESENT

Era/Period		Year	Events

CENOZOIC

CRETACEOUS — Dinosaurs, pterosaurs, ichthyosaurs, plesiosaurs and ammonites

Mammals diversify

CRETACEOUS EXTINCTIONS

— 65

First primates

Placental mammals

Spread of flowering plants

Butterflies, moths and termites appear

— 135

JURASSIC

First birds
Flies

— 190

TRIASSIC

Earliest mammals
First dinosaurs

— 225

PERMIAN

Mammal-like reptiles (therapsids)

Peak of plant diversity

— 280 — **GLACIATION** ∙ Pangaea forms ∙ **PERMIAN EXTINCTIONS**

CARBONIFEROUS

Ants, bees and wasps
Beetles   Reptiles
Complete metamorphosis
Mayflies, bristletails

Foldable wings

— 345 — Simple wings
Seed ferns and conifers

DEVONIAN

Gilboa fossils

First amphibians—
spread of fishes ——— First forests ———

— 395

SILURIAN

Springtails
Millipede-like myriapods
First air breathers
First land plants

— 430 — **GLACIATION** ———

ORDOVICIAN

— 500 — First vertebrates ———

CAMBRIAN

Chitinous exoskeletons

— 570

CENOZOIC

Epoch	Events	Year
PLIOCENE	Human evolution, modern mammals. **PLEISTOCENE**	3
	Apes, mastodons, *Pliohippus* (one toed horse), bear-like dog, rhinoceroses, saber tooth tigers.	5
MIOCENE	great apes, development of grasslands *Parahippus* (three toed horse), wolf like dog, burrowing beaver, pig-like animal. **Dryer climates**	24
OLIGOCENE	Early rhinoceros, ancestral dogs and cats, primitive rodents, ancestral camel, *Brontotherium*. **Colder temperatures**	37
EOCENE	*Titanotheres*, ancestral horses *(Orohippus)*, primitive tapirs, lemur-like monkeys, crocodiles.	58
PALEOCENE		65

Dinosaurs in the Garden
An Evolutionary Guide to Backyard Biology

Dinosaurs in the Garden

An Evolutionary Guide to Backyard Biology

Written and Illustrated by
R. Gary Raham

Plexus Publishing, Inc.
Medford, NJ

Contents

Acknowledgments

With the possible exclusion of diaries, I can think of no books that are written without encouragement and feedback, and this one was no exception. I particularly want to thank Dr. Ross H. Arnett, Jr. at Flora and Fauna Publications for his editing skills, his enthusiasm for my artwork, and his labors at reviewing, and finding reviewers for, many of the chapters. I appreciate the work of Thomas Hogan and his staff at Plexus Publishing, particularly Kay D'Attilio for her editing assistance and Marcia Dobbs for her creative design and layout of the text.

Dr. William Marquardt at Colorado State University gave me advice on Chapters 3 and 4. Professor Howard Crum at the University of Michigan reviewed the material on fungi and lichens in Chapters 5 and 6. Janice Moore at CSU ferreted out problems with the pill bugs discussed in Chapter 9 and the starlings of Chapter 13. Dr. G.B. Edwards, Florida Department of Agriculture, helped with Chapter 10 (spiders), and Professor Barry D. Valentine at Ohio State University reviewed the chapter on salamanders. My thanks to all these individuals for their technical advice. However, I will assume any blame for errors that may have crept in despite their efforts.

Much of the factual content and the illustrations for Chapter 9 (pill bugs) first appeared in "Pill Bug Biology: A Spider's Spinach, But a Biologist's Delight" in the January 1986 issue of the *American Biology Teacher*.

In cases where I have redrawn illustrations or photos from the existing literature I have made every effort to give credit at the end of the captions. The quotations by Leeuwenhoek in Chapter 3 are taken from the Dover edition of Clifford Dobell's book, *Antony Van Leeuwenhoek and His 'Little Animals'*.

Special thanks to the Denver Writer's Group under the tutelage of Ed Bryant, who helped hone my writing skills; to Shirley Parrish who tried to keep my English usage and commas in order in the last few chapters; to Nik Berrong, a supportive fellow naturalist; and to my wife, Sharon, for proofreading and perseverance. Thanks to Deanna and Lindsay for sharing their Dad with a book.

Preface

As you will discover very quickly, this book is not entirely about dinosaurs. At least it doesn't discuss *Tyrannosaurus rex, Stegasaurus,* or any of the other typical inhabitants of the Mesozoic whose names may or may not be household words for you. You can be sure, however, that there are dinosaurs in your garden. Dinosaurs are not as exotic as you may think. Many are alive and well, though not nearly as large as their predecessors. The cast of characters has changed in the last 150 million years, but the acting company that stages the play is the same. We are fortunate to be alive when humans (and other mammals) can play more than a bit part.

I chose *Dinosaurs in the Garden* as a title for this book because I wanted to create an image of the exotic and wonderful close at hand. Rare sights and engaging mysteries are nearby for anyone with an interest in nature. You may not live near a forest or beach or mountain range. Everything you see out your window may seem manmade. But even in places where nature seems remote, it isn't. You need only look in the right places and adjust your perspective to encompass some unfamiliar dimensions.

Many times in science classes we learn of wondrous creatures we may only see in zoos, television documentaries, or on a retirement cruise from the deck of a ship. That's fine. But we shouldn't lose the opportunity to appreciate the amazing collection of plants and animals that are as near as our backyard and how they fit into the historical drams of evolution that led to those more "ostentatious" forms.

Science teachers were particularly on my mind as I wrote this book, most likely because I have been one and would have valued a book that could stimulate my curiosity, provide me with ideas for classroom demonstrations and experiments, and direct me to further sources of information. I hope this book accomplishes those ends as well as just being a "good read" for anyone interested in natural history.

1

Taking a Look Around

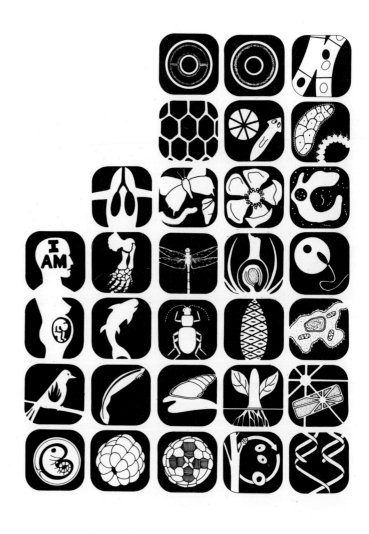

Don't get too comfortable yet.

I do appreciate the fact that you're taking the time to read *Dinosaurs in the Garden*, but I'd like you to do two things before you settle into your easy chair. First, take a moment to look out of the nearest window. If you're at home, chances are you've seen the view many times before. Perhaps you're looking at a well-kept yard in the suburbs with close-cropped grass and stately shade trees. Perhaps you're looking at farmland with cultivated fields stretching to the horizon. Or perhaps your view is filled by man-made things: flashing signs, tall buildings, and bustling vehicles. In any case you're not seeing all there is to see. Your backyard, wherever you might be, is home to many living creatures besides yourself—including, as we will see later, a few dinosaurs in your garden. They've found your backyard a likable place, and many live their lives there without your knowledge. Some of these creatures are too small to see, floating in the air or hiding in the soil. Others are out when you're not, going about their business in the quiet, dark hours of the night. Still others remain unseen simply because you're not looking for them.

The second thing I'd like you to do is go outside. Find a large rock, a cement culvert beneath a drainspout, or a board that has been lying on the ground or against a wall for a long time. Before you turn over that

Figure 1-1: *To see some aspects of the natural world around you, you have to adjust your perspective. To appreciate the interrelatedness of the living world and its historical roots, it helps to engage your imagination.*

rock or lift up that board, however, prepare yourself to remember the various things you will see. When you're ready, go ahead. When you think you've seen all there is to see, carefully replace the rock, culvert, or board the way it was.

Now you can get comfortable and make a list of what you've seen. The variety of creatures you've discovered will depend

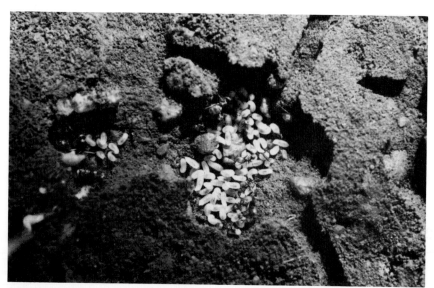

Figure 1-2: Lifting a rock or board in ant country will cause the adults to rapidly move all their pupae to underground chambers. Replace disturbed rocks etc. after making your observations.

partly on the place you live and the time of year, but there will also be many creatures that can appear in widely different places. Compare your list with mine, which was made after turning over the cement slab under a downspout in a suburb of a small, north central Colorado town:

Crickets
Ants (they had an extensive tunnel network and were scurrying around trying to get their immature stages, which look like white oblong beans, to deeper parts of the tunnels)
Aphids (very tiny insects which ants sometimes keep with them in their tunnels and "milk" for the sweet drops they produce from their rear ends)
Spiders or "daddy longlegs"
Mites (small relatives of the spiders that also have eight legs)
Earthworms
Pill bugs (they look like little armadillos and roll up into a ball when you touch them; they are related to crayfish)

Centipedes or millipedes (worm-like in shape, and segmented; centipedes have two legs per segment whereas millipedes have four shorter legs per segment)
Slug (a shell-less snail)
A green stain on part of the soil which represents a colony of single celled plants

If I had dug up part of the soil and examined it with a hand lens or microscope, I could have found much more: round worms, protozoans, plant spores, smaller mites, eggs of various insects or their larvae, rotifers . . . the list could go on. And these are the critters in or on only one square foot of soil. Other collections of organisms can be found in the trees or in the garden or floating in the air we breathe. Every kind is unique, and every kind has carved out a living that is somehow dependent on the other creatures around it—including you.

A host of animals and plants have adapted quite nicely to conditions in your backyard, just as they have adapted to much harsher conditions elsewhere. Consider, for example, conditions on the bottom of a deep-ocean trench. There, in near total darkness, eight-foot worms, eyeless, with no mouths or digestive tracts, weave back and forth in the water. The worms live at the edge of a vent on the ocean floor. Geysers of boiling water burst from the vent, carrying with them rich sprays of minerals from inside the Earth. When the water from these vents hits the cold, ocean-bottom water, compounds of sulfur, iron, zinc, and other elements rain down on the surrounding creatures. For the worms, giant clams, mussels, limpets, and crabs that live here, the rain of chemicals is as important to them as the sun is for most animals. Specialized bacteria, living symbiotically with many of these creatures, break down sulfur or iron compounds and release energy which is used by their hosts and ultimately passed to all other members of this living community.

Or consider the Antarctic, where winds whistle through dry valleys and cut the sandstone into raw-edged shapes. The temperature in summer may reach 30°F, but it can drop to a low of -158°F in winter. In some respects these valleys are nearly as unfriendly to life as the surface of Mars. Yet, if you break some of the ragged rocks, as scientists first did in 1974, you can find life inside. Lichens, composite organisms made up of simple green plants called algae and non-green plants called fungi, live inside the rock just beneath the surface. The fungus dissolves minerals from the rock, which the green algae use with sunlight to make sugars. Natural fertilizer falls from the sky with the snow when lightning provides the energy for some of the nitrogen and oxygen in the air to combine and form nitrates.

In the tropical forests of the world where it is warm and wet and there are few extremes in temperature or rain, life forms grow and multiply in bewildering numbers and kinds. In a quarter acre of rain forest you can count over 200 species of trees. Fifty species of plants and animals may live on each tree. Over two-thirds of the four to ten million kinds of plants and animals that share our world live here, where weather is kind but competition is fierce.

Living things can meet almost every challenge the Earth has to offer. A dramatic example of this occurred one hundred years ago in the East Indies. In 1883 a volcano on the island of Krakatoa exploded with the violence of a hydrogen bomb. Six cubic miles of rock were thrown into the air. An ash-covered peak was all that remained of the island. So much debris was pumped into the atmosphere that sunsets were effected worldwide for days, prompting Tennyson to write: "For day by day, thro' many a blood-red eve . . . The wrathful sunset glared." But within nine months, life had returned to Krakatoa. A spider landed on the barren island after floating at least 25 miles over the ocean beneath a delicate parachute of silk. Spores fell on the island. Seeds were washed ashore on floating debris or dropped in the waste of visiting birds. In three years there were 11 species of ferns and 15 kinds of flowering plants. In ten years there were coconut tree seedlings, sugar cane, and orchids. In less than fifty years the island was covered with dense forest and held 47 species of birds, bats, rats, reptiles, and hundreds of kinds of insects. Today, as beautifully documented by Dieter and Mary Plage (see references), Krakatoa and its neighboring islands are sanctuaries for a rich assortment of plant and animal life.

Where did all this life come from? This is something of a trick question, because there is more than one way to answer it. You may think we've wandered some distance from your backyard to even ask the question, but we haven't. In fact, we can ask the question about the life in your backyard as easily as we can about some distant island. Where did the ants, the crickets, and the spiders come from?

"They've always been there" is the answer that most easily comes to mind, but their tenure there could be relatively recent. Before your house was built there was probably a field or farm where you're standing now. Before that, there was most likely open prairie or forest depending on where you live. And before that, ten thousand years ago or so, if you live north of Ohio, there may have been a towering wall of ice where you now stand, chill winds blowing along its flanks and ruffling the shaggy hair on grazing mammoths. Go back farther still and there may have been an inland sea covering your backyard with clams and strange worms where petunias now grow. Obviously, it takes different kinds of creatures to live in these varied conditions, and the animals and plants now in your yard moved in when they could make a living there. This is one answer to the question of where your animal and plant neighbors came from. Finding out how they got where they did and how they get along with each other (and with you) is part of what *Dinosaurs in the Garden* is about.

When you ask the question "Where did this critter come from?" you are also asking about the history of life on Earth. And that, too, is part of what *Dinosaurs in the Garden* is about because it deals with the changes that have occurred in organisms over the length of Earth's long history. The changes in organisms through time, or evolution, is family history taken to the extreme. Although *Tyrannosaurus* no longer makes the earth tremble with his thundering stride, fragments of what he was linger on in the strut of a squawking starling. Evolutionary concepts are attempts to make sense of the similarities found in all life on Earth, and as such they are integral parts of modern biology.

Mostly, however, *Dinosaurs in the Garden* is an introduction to a few common creatures that will, I hope, prime your enthusiasm for further discovery. To that end, this book offers suggestions for observations and experiments, along with appendices of reference material and supplementary information for those who are interested. For example, "Notes for a Prospective Naturalist," which starts on page 8, contains hints on how to get started as a naturalist and some things to consider as you begin your explorations.

Over the long span of time the Earth has existed, living things have had to adapt to many changes: the changing temperature of the Earth as it cooled, the changing composition of the atmosphere, the changing environments caused by shifting continents, and the changes that occurred in other creatures that made them more efficient. Certain changes were more successful than others and resulted in groups of organisms that dominated the Earth of their time. Many of the creatures in your backyard display these major innovations.

As you proceed in this book you will see certain symbols at the beginning of new sections. These symbols represent some of life's biggest "inventions." I hope the symbols will simplify some concepts without doing an injustice to the real complexity of the living world. Chapter 2 summarizes these biological concepts and describes one modern scheme of classification. You may want to skim some of this information the first time through and refer back to it as individual chapters deal with specific examples. Chapter 15 provides historical background about how naturalists' attitudes and ideas about nature have changed.

All the fascination, mystery, and wonder that you could hope for is all around you. The creatures common to my backyard may not always be the same as those that inhabit yours, but that's the fun of it. Look at your backyards closely. They are a jungle of ongoing experiments in how to adapt to a changing planet. The mysterious mix of creatures that exist there normally live, and die, beneath your attention. But if you look carefully, with an open mind and lots of curiosity, you can follow these creatures around the trees, through the garden, and even back through time.

Notes for a Prospective Naturalist

If you are relatively new at the study of nature, you may appreciate a few tips on getting started. Supplies for a day-pack, proper notebooks, and a respectful attitude are all important for the naturalist.

Stocking the Day-pack

Some of the things you take on any given expedition will depend on where you go, of course, but many of the items below will be generally useful:

Hand lens (these are usually about ten power magnification)
Tweezers
Thread
Nets (see Chapter 4 for aquatic collecting)
Pocket knife
Notebook and pen (see the next section, starting on page 9)
Field guides
Binoculars
Camera
Containers of assorted sizes
Envelopes, "sandwich" bags, waxpaper
Tape
Local and regional maps (topographical maps of virtually any spot in the U.S. can be obtained from the U.S. Geological Survey, Denver, Colorado 80225 or Washington, D.C. 20242. They will send a descriptive brochure on what is available on request.)
Killing jar (for insects). These can be made from a large wide-mouth

jar (peanut butter jars are good). Fill the bottom of the jar with about an inch and a half of plaster of paris. Add a small quantity of killing agent, usually ethyl acetate, available from chemical supply companies, and place a few shreds of paper toweling in the bottom to keep the insects from direct contact with the poison. The jar will have to be recharged every now and then with poison. **Insect aspirator**. These can be purchased from biological supply companies or made. Take a clear plastic vial and cut out the bottom. Put corks that have a hole big enough to accept glass tubing in each end of the vial. Put short pieces of tubing through each cork. Attach a length of rubber tubing to one end. You can collect small insects and other creatures by sucking them into the glass vial (see Figure 1-3).

A camera is a particularly good way to "collect" animals and plants without removing them from their habitats. Thirty-five-mm SLR cameras offer considerable versatility. Most brands offer "systems" of attachments, such as lenses of different focal lengths, flashes, auto-wind features, and filters. If you are at all interested in

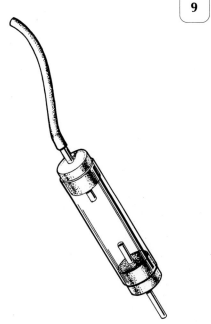

Figure 1-3

photography, it makes a good adjunct to nature watching. I have found Olympus products to be excellent and lightweight. They also manufacture microscopes and other optical products and have interfaces and accessories for microphotography.

Notebooks

Properly kept notebooks are crucial for getting any valuable long-term information from your nature studies. You should have two types: a field notebook and a permanent journal.

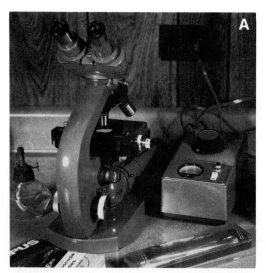 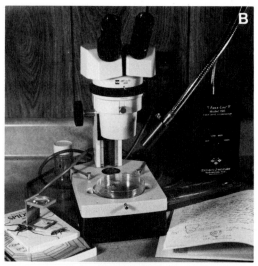

Figures 1-4A & 4B: *Compound microscopes (**A**) are best for studies of protozoa, bacteria and thin sections of higher organisms. Complexity and prices of microscopes vary greatly. The most important aspect is the quality of the optics. If there are rainbow colors around the image, for example, there is strong chromatic abberation in the lens, which is not good. A model like this one, with binocular head, oil immersion lens, mechanical stage and transformer would probably cost $1500 or so when new.*

*Dissecting scopes (**B**) usually magnify from ten to forty times. They are good for insects, fungi, and a host of small creatures that are too large and opaque for a microscope. Lighting can come through the base and/or from the sides, allowing you greater options for viewing specimens. As an all-around tool the dissecting scope is probably more useful to the naturalist than the microscope. Although this model is worth about $1000, you can get quite serviceable models for several hundred dollars.*

A good field notebook is a bound, stiff covered book with blank pages. You can also get them with rules or gridded pages, but the blank pages are much better for sketches. All naturalists, whether they consider themselves artists or not, should be prepared to include many sketches in these books. Notes should be taken on the spot with this notebook, as memories are exceptionally fallible. Consider the five "W"s when recording in a field notebook: Where, When, What, Why and With whom or what?

"Where" includes county, state, and specific locality. It should also include something about the specific habitat (i.e. on the north side of a pinyon pine tree, and beneath a fallen log).

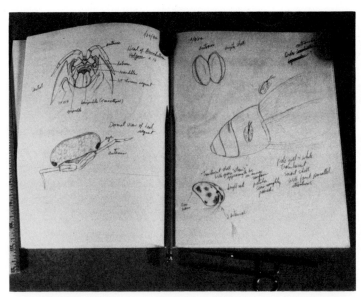

Figure 1-5: Field notebooks will not always be neat, but should contain the five W's: Where, When, What, Why and With whom or what. Without verifying information regarding time of year, locality and so forth much of the value of your observations may be lost.

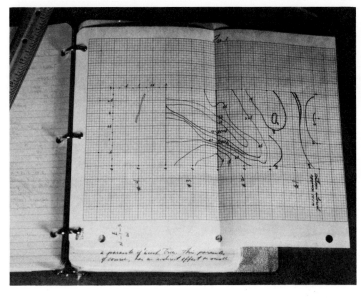

Figure 1-6: Permanent notebooks should be kept up to date as quickly as possible and should contain a more detailed analysis of your observations with appropriate summaries, charts and/or mathematical manipulations. You may also want to include photos.

"When" should cover the day and date as well as time of day.

"What" should include a description of what you found and observations on behavior, coloration, etc.

"Why" can cover speculations on certain behavior patterns, reasons for pest damage and so forth. You will undoubtedly reconsider the "whys" later, but on-the-spot hunches are often good ones.

"With whom or what" pertains to observations on the animals and plants associated with your find as well as notes on

A Code of Conduct

It is up to every one of us to minimize our impact on nature, even as we participate in and enjoy it. The following list of guidelines gives you a few things to keep in mind as you discover the natural world.

1. Take no more specimens than absolutely necessary and avoid overworking a particular area. Take photos when possible or capture and then release organisms.

2. Don't destroy all the potential habitats of an organism in your searches. Don't, for example, turn over all the logs in an area when searching for salamanders. Return disturbed logs, stones, etc. to the way you found them.

3. Cut, don't break, plant samples. Breaks will be more susceptible to infection and further injury.

4. Avoid rare species, and be aware of and obey local laws protecting certain species.

5. Leave things as you found them. Litter and disturbed habitat, besides being unpleasant for others, may give predators an unfair advantage in locating their prey.

6. Always get permission for collecting on private land.

7. Don't overdo baiting. Animals become used to the extra source of food.

8. Avoid poisons, drugs, and sticky materials that may persist in the environment and affect many organisms.

9. Consider the welfare of captive animals first, before the goals of your project.

10. Don't share the location of collecting areas with people who may abuse the habitat.

the weather, temperature, and other factors. Information that may have seemed irrelevant at the time can prove very valuable later.

Drawings are particularly useful for summarizing observations with a minimum of words. A small selection of colored pencils allows you to indicate colors.

When you get home you should transfer and coordinate your information into a permanent notebook. Loose-leaf notebooks work best for this, because you can add or subtract pages as necessary and add graphs, charts, maps, and so forth. A 7 x 10 inch size works well for me, but any size that is convenient will do. Number all pages, and as information accumulates, index the notebook. If you can't easily locate something when you want it, it's not of much permanent value. Some people also use a file card system arranged by topic or organism with references to journal entries and perhaps relevant literature citations.

REFERENCES

Farb, Peter. 1963. *Ecology*. New York: Time Incorporated.

Plage, Dieter and Mary. 1985. Return of Java's Wildlife. *National Geographic,* vol. 167, no. 6 (June).

Science News staff. 1985. Animals at the Hydrothermal Vents. *Science News,* vol. 128, no. 8 (August 24).

Science News staff. 1977. Living Sea: Life on the Galapogos Rift. *Science News,* vol. 111, no. 18 (April 30).

Science News staff. 1981. Tube Worm Nourished With Help From Within. *Science News,* vol. 120, no. 3 (July 18).

White, Peter T. 1983. Nature's Dwindling Treasures: Rain Forests. *National Geographic,* vol. 163, no. 1 (January).

2

Surveying the Kingdoms

Consider this chapter as a roadmap to your study of nature. If you already know the terrain quite well, feel free to skip it and jump ahead to other chapters. If not, you may want to take a look at one modern classification scheme presented here to see how the various parts of the living world fit together. You will also find a description of "innovations," which is my personal list of key events in the history of life that seem to have determined major branching points in the evolutionary tree. You may want to refer back to these definitions as concepts are discussed in individual chapters. Keep in mind that you will be introduced to only a very few animals and plants in this book, but as you discover new ones on your own, consider how they fit into the five living kingdoms.

The Five Kingdom System

The four to ten million estimated species that share our planet form a bewildering network of organisms that must live with each other and with the changing conditions of the Earth. Although classification systems are artificial, they do help to get some grasp of what's going on. A modern system based on the work of R.H. Whittaker (1924–1980) is very useful because it incorporates much of the information accumulated in the twentieth century about viruses, bacteria, and fungi. An excellent reference is *Five Kingdoms, An Illustrated Guide to the Phyla of Life on Earth* by Lynn Margulis and Karlene V. Schwartz.

The five kingdoms are:

MONERA: This includes bacteria and cyanobacteria. The latter used to be called blue-green algae. Sixteen phyla are recognized. Chapter 3, "Microscopic Gardens," describes how one man's curiosity and persistence uncovered the existence of these creatures, and how you can rediscover them for yourself.

PROTOCTISTA: This kingdom includes protozoans, algae, and some organisms, like cellular slime molds, that used to be classified with the fungi. Organisms in this group and all higher groups have a more complex cell structure than the Monera. Organization is basically at the single-cell level rather than multicellular construction with different cells specialized for different functions. Twenty-seven phyla are recognized. Chapter 4 gives you a glimpse into the variety and complexity of this group.

FUNGI: Whittaker recognized that fungi are basically different than either plants or animals. Five phyla are recognized. Chapter 5 looks at the world from a mushroom's point of view. You will also find that fungi, despite their immobility, are successful predators in their minute habitats.

ANIMALIA: A wealth of animals are recognized, divided into 32 phyla. Chapters 7–14 deal with various representatives of this kingdom.

PLANTAE: There are nine plant phyla. Most that you encounter in your backyards and elsewhere belong to either the flowering plants (Angiosperms) or conifers (Gymnosperms). Mosses (Bryophytes) and ferns (Filicinophyta) are also fairly common, however. The other five phyla represent remnant species that were much more important members of living communities in the past. Chapter 6 deals with the unique liaison between algae and fungi to form lichens. Chapters 11 and 14 discuss "higher" plants and their shifting relationships with the various members of the animal kingdom through time. You are invited to observe some of those interactions in your own backyard.

Figure 2-1 shows how these major groups are believed to relate to each other. Protoctistans developed from Moneran ancestors, and Fungi, Animals, and Plants were Protoctistan experiments.

Nature's Innovations

The interrelationships among living things are often portrayed in the form of a family tree with all life originating from a common ancestor. Although many experiments may have occurred in the flux of energies and raw material of our planet's youth, several facts point to the conclusion that a single type of living organism gave rise to all the living creatures we see today. All life on Earth, for example (with the possible exception of certain protozoans), has the same genetic code. That's why bacteria can successfully produce human hormones when the proper gene is spliced into their genetic material. All biologically active proteins (enzymes, hormones, etc.) in all living things are "left-handed." There are "right-handed" proteins that are chemically identical, but are mirror images of their counterparts. These are functionless. The implication is that left-handed chemicals, by chance, initially were part of the original lineage, and the pattern was followed from there on out. Also, the fossil record supports a general trend from the simple to the more complex; the "new" species that appear in successive geological eras have structural ties with the forms that preceded them.

Life is by nature conservative. When something works well there is great value

The Precambrian Roots
of the Five Kingdoms

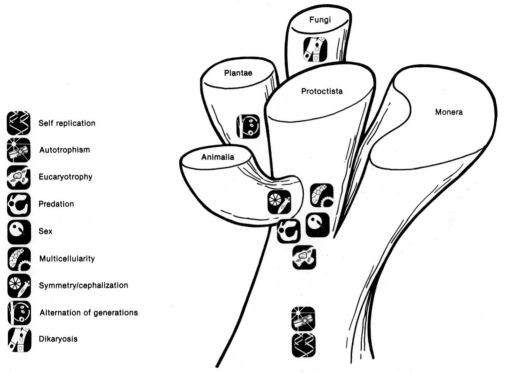

Self replication

Autotrophism

Eucaryotrophy

Predation

Sex

Multicellularity

Symmetry/cephalization

Alternation of generations

Dikaryosis

Fungi

Plantae

Protoctista

Monera

Animalia

Figure 2-1: The ability of complex molecules to make copies of themselves and the harnessing of chemical and electromagnetic energy sources paved the way for more complex organisms. All kingdoms except Monera have a complex cell structure with an organized nucleus and intercellular organelles. Predation, multicellularity and sexual reproduction allowed protoctistans to flourish. Plants, animals, and fungi developed from the protoctista.

in retaining it. Therefore, a number of mechanisms have evolved to repair damage to genetic material to insure changes don't occur. For billions of years bacteria were the highest forms of life and did quite well at trapping solar or chemical energy to make more bacteria. When a major innovation comes along, however, new paths or branches are created. Often the old ones remain too, perhaps with more restricted ranges than they once had. Each time a change occurs, the possibilities for further changes are multiplied. Thus, a tree-like pattern does emerge with the major innovations marking the branching points.

What I will proceed to do in *Dinosaurs in the Garden* is to identify key innovations that occurred in the history of life, and code them with symbols. As you deal with individual organisms around you in the separate chapters, you can glance at the symbols at the beginning of a chapter and refer back to some of the charts here to see how those organisms fit into the overall pattern.

Here is a synopsis of this symbolic check list:

 SELF REPLICATION: The DNA molecule (deoxyribonucleic acid) or its close relative RNA (ribonucleic acid) is the basis of reproduction for all Earth life. Its "unzippering" spiral form is symbolized. Without the ability to self-replicate, living forms could not "accumulate" hereditary changes.

 AUTOTROPHISM: Very early in the history of life, techniques developed for using sunlight or chemical sources of energy to make food.

 EUKARYOTROPHY: The first branch in life's tree resulted when the simple cell structure of bacterial forms developed into the more complicated type found in all higher organisms. Bacterial cells are called prokaryotes, and all other cells are known as eukaryotes. The differences are quite profound, as you will see in Chapter 4. Scientists currently believe that the primitive eukaryotic cells may have been a parasitic or symbiotic relationship between two or more prokaryotes.

 PREDATION: The development of cells that eat other cells rather than make their own food is thought to be one of the factors that accelerated the rate of evolution. The decimation of the vast bacterial gardens opened up opportunities for the development of new producers and consumers.

 SEX: True sexual systems first appear among the "new" eukaryotic cells. Bacteria have a type of genetic exchange, but it is irregular and not linked with cell reproduction. Sex allows for genetic diversity from one generation to the next and thus probably contributed to the rapid acceleration of evolution.

MULTICELLULARITY AND DIFFERENTIATION: The development of higher organisms was dependent on cells "sticking together" into multi-celled colonies. Ultimately, different cells specialized in different functions, losing the ability to survive independently. Collections of cells specializing in a certain function are tissues.

 SYMMETRY/CEPHALIZATION: Animals are mobile plant eaters. Because they can't make their own food, they are heterotrophs rather than autotrophs. The development of symmetry, and a head and a tail end, the former with a concentration of nervous tissue to sense and interact with the environment, are innovations that led to the animal kingdom.

 DIKARYOSIS: Fungi are also plant and detritus eaters, that is, heterotrophs. For the most part, however, they "stay put" in their environment. A distinctive feature of fungi is dikaryosis. Fungal filaments fuse, and nuclei of different individuals can coexist in the same filament for long periods. Eventually the nuclei can fuse, and sexual recombination can occur. Figure 2-2 shows the three major paths fungi have pursued. Zygomycotes are simple molds, like the black bread mold, with spores borne in globes called sporangia on the tips of long stalks. Ascomycotes bear spores in sac-like structures called asci. Molds, yeast and most lichen fungi are ascomycotes. Basidiomycotes bear spores on club-shaped basidia (see Chapter 5). Rusts, smuts, and mushrooms belong to this group.

 ALTERNATION OF GENERATIONS: Alternation of generations is a characteristic usually associated with plants, although it results from the need to maintain a fixed amount of genetic information in a species while still enjoying the recombinational advantages of sex. A generation with a single set of chromosomes (haploids) alternates with a generation with a double set of chromosomes (diploids). In animals the haploid generation is represented by the sex cells, eggs and

Fungal Relationships

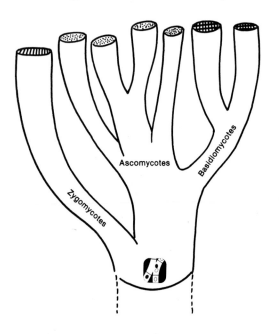

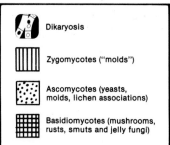

Dikaryosis

Zygomycotes ("molds")

Ascomycotes (yeasts, molds, lichen associations)

Basidiomycotes (mushrooms, rusts, smuts and jelly fungi)

Figure 2-2: Dikaryosis, or cell organization where nuclei from two strains coexist without fusion for long periods, is characteristic of all fungi. The three major divisions of fungi are distinguished by the form and complexity of their reproductive structures.

sperm. In primitive plants both haploids and diploids can be free-living and may look quite different. Haploids are also called gametophytes because they give rise to the gametes, or sex cells. The sex cells unite to form the diploid generation, called sporophytes. The sporophytes produce spores that germinate into the gametophytes. Higher plants are virtually all sporophytes. The gametophyte stage has become reduced, shortened, and the entire stage takes place within the sporophyte plant body.

Innovations Along the Plant Branch
(see Figure 2-3)

VASCULAR TISSUE: Vascular tissue is comprised of cells specialized for carrying water and food to all parts of the plant. Without it the size to which a land plant can grow is limited. Mosses and liverworts lack this feature; all higher plants retain it.

SEEDS: When fern-like plants in the Carboniferous retained the gametophyte generation within protective bracts of the sporophyte, a new milestone was reached in plant development. The survival of the gametophyte was no longer so risky.

 ENCLOSED SEEDS: The seeds of angiosperms are covered by protective tissue. In contrast, gymnosperms have "naked seeds." Angiosperms also possess double fertilization of the female gametophyte, which usually leads to the production of a food source for the young plant called the endosperm. The endosperm actually makes up most of the volume of a seed.

 FLOWERS: About 138 million years ago, at the beginning of the Cretaceous, flowers developed among the angiosperms. Flowers resulted in a symbiosis with the animal kingdom that helped insure more efficient fertilization, and thus reproduction. Flowering plants dominate the world's flora today.

Innovations Along the Animal Branch
(see Figure 2-4)

 COELOM: Very simple animals, like sponges and jellyfish, have only two kinds of tissue: outside tissue (ectoderm) and inside tissue (endoderm). More advanced forms, like flatworms, have a "middle tissue," or mesoderm, from which such things as muscles develop. Within this middle tissue a cavity, called the coelom, develops in all higher animals. Various internal organs develop within the coelom.

 PSEUDOCOELOM: Some animals, broadly grouped in my . diagram as "worm phyla," have a "quasi-coelom" that is not surrounded by mesoderm. This group includes the "rotor-headed" animals (rotifers) commonly seen under the microscope.

 PROTEROSTOMES: Among the animals with coeloms, one major branch that includes mollusks and arthropods has the coelom derived from

Innovations Among the Plants

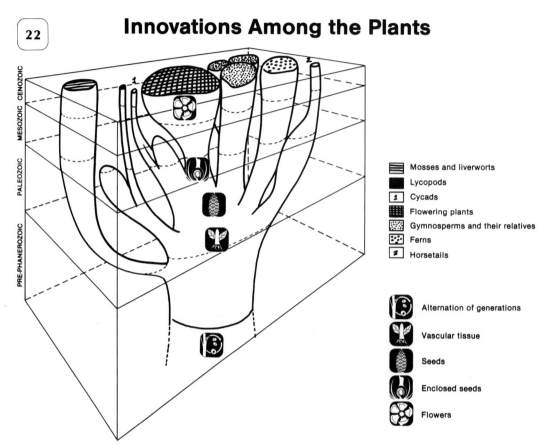

Figure 2-3: *Plant evolution has involved adaptation to terrestrial environments and more efficient and effective ways to reproduce. Flowering plants represent the most recent refinements in the plant kingdom.*

spaces within the mesoderm.

DEUTEROSTOMES: Starfish and chordates (including the vertebrates) took another route. The coelom is formed from outpocketings from the gut, which is made of endodermal tissue.

EXTERNAL SKELETON: If we follow the proterostome side of the animal kingdom for a while, we find that they utilized excess minerals to create an external skeleton. Without a skeleton of some kind the size which animals can reach is limited.

Innovations Along the Animal Branch

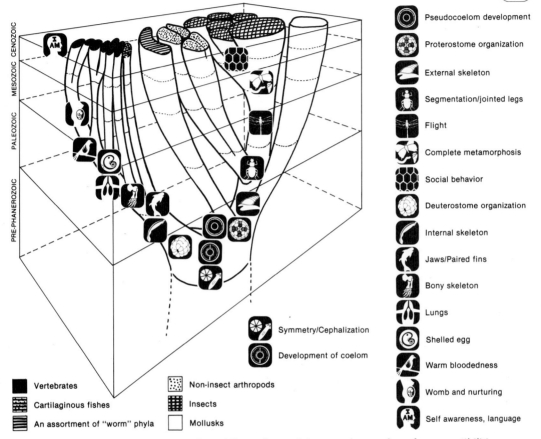

Pseudocoelom development

Proterostome organization

External skeleton

Segmentation/jointed legs

Flight

Complete metamorphosis

Social behavior

Deuterostome organization

Internal skeleton

Jaws/Paired fins

Bony skeleton

Lungs

Shelled egg

Warm bloodedness

Womb and nurturing

Self awareness, language

Symmetry/Cephalization

Development of coelom

Vertebrates

Cartilaginous fishes

An assortment of "worm" phyla

Non-insect arthropods

Insects

Mollusks

Figure 2-4: *As "innovations" continue along different lines of descent, the number of new possibilities increases. Animal life took its major fork when one branch explored the external skeleton and the other developed the internal skeleton. In terms of sheer numbers, the invertebrate insects came out on top—however, humans take pride in their large brains and clever hands.*

 SEGMENTATION/JOINTED LEGS: While mollusks were exploring the variations possible with the shell, arthropods tried out the possibilities of segmentation and jointed appendages.

 FLIGHT: Flight opened up a whole new environment for insects, a group which dominates the Earth today in terms of numbers and kinds.

Body Cavities

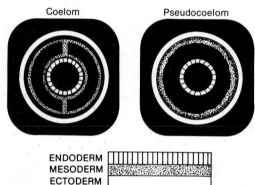

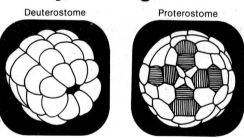

Figure 2-5: *The organization of the three basic kinds of tissue which make up animals is fundamental. Animals with body cavities lined with mesoderm (coelomates) have been the most successful. Some simple organisms have a body cavity that forms between the mesoderm and endoderm. Such a cavity is called a pseudocoelom.*

Embryonic Organization

Figure 2-6: *Embryonic development is distinct between vertebrates and invertebrates. In Deuterostomes (the vertebrate branch), the spot on the early embryo that indents to become the inner tissues later becomes the anus. In Proterostomes this spot becomes the mouth. The "annelid cross" shown in the proterostome embryo is mesodermal tissue which is destined to form the coelom.*

COMPLETE METAMORPHOSIS: Complete metamorphosis, a characteristic of such insect groups as the beetles and butterflies, allowed adult and immature forms to exploit different niches in their environment and thus not compete with each other for food and space. The adult form also became a specialist in reproduction. There are more kinds of beetles than all other animals put together—a testimony to their success.

SOCIAL BEHAVIOR: Insect groups like the bees and ants developed complex social systems built on instinctual behavior that have made them, too, extremely successful.

INTERNAL SKELETON: The deuterostome branch of the animal kingdom eventually developed an internal skeleton. The main body nerve runs dorsally (along the back) near a cartilaginous rod called the notochord.

JAWS/PAIRED FINS: The development of jaws and paired fins some 400 million years ago gave early types of fish an edge over some other marine predators, such as the arthropod *Eurypterus*. Jaws gave offensive capability and the paired fins provided greater swimming agility.

 BONY SKELETON: The development of the bony skeleton, along with such things as the swim bladder and lateral line sensory system, gave bony fish an advantage in many habitats and preadapted them for a later assault on the land.

Figure 2-7: Shelled mollusks such as these snails (Turritella sp.) from Wyoming are common fossils. These were freshwater animals that lived during the Eocene. The development of shells was a major innovation in the history of life.

 LUNGS: The modification of the swim bladder as an accessory breathing device, together with the modification of the two sets of paired fins as legs, allowed amphibians to develop and exploit a land habitat.

 SHELLED EGG: Development of the shelled egg allowed reptiles to be completely independent of water, even for reproduction.

 WARM BLOODEDNESS (Homeothermy): The ability to maintain a constant body temperature required burning more calories, but resulted in quick, active animals that were not as dependent on the time of day or seasonal variations in weather. Current research indicates that at least some dinosaurs were warm blooded. Their descendents became birds. Mammals were also recipients of this ability via mammal-like reptiles called therapsids.

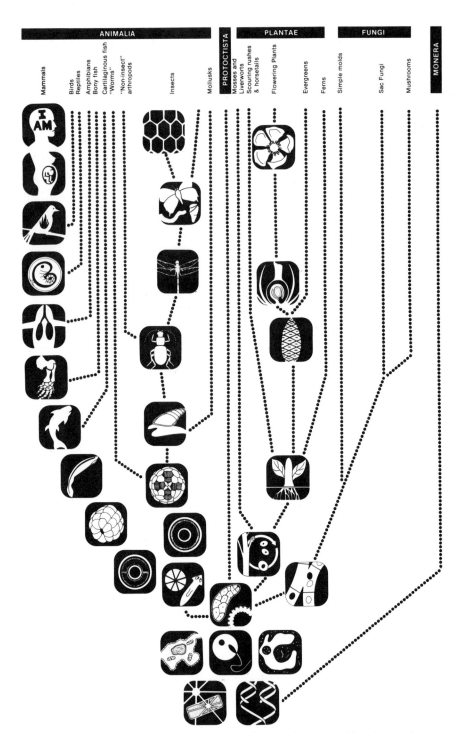

Figure 2-8: *The major phyla of living things are displayed in terms of key "innovations" that resulted in distinct lines of descent. Such innovations don't necessarily result in the death of predecessor species, but may limit or change their ranges.*

WOMB AND NURTURING: Mammals moved away from the tendency to have large numbers of young which had to fend for themselves. Embryos are few but are retained within the mother's body and derive nourishment directly from her. After birth, young mammals are fed from milk glands, and one or both parents invest much time in training them.

SELF-AWARENESS/LAN-GUAGE: The development of mammals involved an increasingly greater brain size. Learning, rather than instinctual behavior, assumed a greater role. In Man this culminated in self-awareness and language. Language has allowed the accumulation of group learning, a sort of cultural heredity, that has greatly accelerated his mastery of the environment. Self-awareness has also given man awareness of his power over other parts of nature and, one hopes, a responsible attitude regarding that power.

Although this list of innovations necessarily simplifies the nature of Nature, it does complement the overall picture. I've omitted detailed sketches of the Moneran and Protoctistan kingdoms because most of what you'll study as backyard biologists will be larger forms, but their stories are fascinating, too, and I hope you'll find the time to explore them in greater detail.

REFERENCES

Margulis, Lynn and Karlene V. Schwartz. 1982. *Five Kingdoms: An Illustrated Guide to the Phyla of Life on Earth.* San Francisco: W.H. Freeman and Company. A very visual reference (black-and-white photos, scanning electron micrographs, and line drawings) that gives capsule descriptions and examples for all phyla.

Meglitsch, Paul A. 1967. *Invertebrate Zoology.* New York: Oxford University Press. This edition has some advantage over later, condensed versions, especially with regards to supplementary material in the appendices.

3

Microscopic Gardens

Discovering a New World

Delft, Holland, 7 September, 1674

"About two hours distant from this Town there lies an inland lake, called the Berkelse Mere, whose bottom in many places is very marshy, or boggy. Its water is in winter very clear, but at the beginning or in the middle of summer it becomes whitish, and there are then little green clouds floating through it: which, according to the saying of the country folk dwelling thereabout, is caused by the dew, which happens to fall at that time, and which they call honey-dew. This water is abounding in fish, which is very good and savoury. Passing just lately over this lake, at a time when the wind blew pretty hard, and seeing the water as above described, I took up a little of it in a glass phial; and examining this water next day, I found floating therein divers earthy particles, and some green streaks, spirally wound serpent-wise, and orderly arranged, after the manner of the copper or tin worms, which distillers use to cool their liquors as they distil over......Others were somewhat longer than an oval, and these were very slow a-moving, and few in number. These animalcules had divers colours, some being whitish and transparent; others with green and very glittering little scales; others again were green in the middle, and before and behind white; others yet were ashen grey. And the motion of most of these animalcules in the water was so swift, and so various, upwards, downwards, and round about, that 'twas wonderful to see...."

Antony van Leeuwenhoek wrote these words three hundred and eleven years ago. He was the first person to see, and recognize as living forms, the creatures we call microorganisms. The serpent-like green streaks were probably strands of the green alga *Spirogyra*. Some of his "whitish and transparent animalcules" were undoubtedly protozoans. Later on he speaks of more little animals that were most likely bacteria: "... and I imagine that ten hundred thousand of these very little animalcules are not so big as an ordinary sand-grain...."

Antony was probably a bit nervous writing his letter, because he was a self-educated man, not particularly comfortable with the written word, writing to Henry Oldenburg, the secretary of the newly formed "Royal Society of London for Improving Natural Knowledge." Natural history was a hobby for Antony, who was primarily a clothes-maker. At various times, however, he was also a surveyor, a wine gauger for the city of Delft, a janitorial supervisor for the town hall, and a general district supervisor.

Curiosity gave Antony his place in history as well as life-long pursuits that served him well over his ninety years. He studied navigation, astronomy, mathematics, nat-

Figure 3-1: Bacteria are some of the smallest creatures on our planet. Approximately 200 would have to be placed end to end to make a string 1 millimeter long. Yet, even they have parasites. The blimp-like structures in this drawing represent bacterial cells called E. coli that live in the human gut. They are being attacked by spidery-looking viruses that inject their genetic material (DNA or RNA, depending on the virus) into the bacteria. The virus RNA merges with the bacterial DNA and directs the bacterial cell to produce viral instead of bacterial protein.

ural and physical science, and biology. Somewhere along the line he picked up the skill of lens grinding, and this ultimately led him to his discovery of microscopic life. His patient work produced biconvex lenses of excellent quality—some magnified over 200 diameters. He mounted these lenses between two metal plates (usually brass or copper—some-times silver or gold) and designed a clever system for moving a specimen into focus in front of the lens. Leeuwenhoek also developed a special viewing technique that apparently allowed him to see his "animalcules" as white spheres against a black field. His craftsmanship and skill gave him a view of the microscopic world that would be unsurpassed for a hundred years.

Today we have optical equipment that makes binocular viewing of bacteria and their relatives routine. Electron microscopes allow us to see images of viruses and even individual large molecules. Yet Leeuwenhoek was the first person to see life which is invisible to the naked eye. This took more than the invention of a good lens. It took a willingness to observe with an open mind—with no preconceived notions of what could or could not be. If you can observe nature in that way, as if you were newly born from one moment to the next, you will not only find a continuing source of enjoyment, but could discover, like Leeuwenhoek, that there is an unimagined universe of living creatures out there.

With or without a microscope you can learn a great deal about the microscopic world. Why not start by making a garden of microorganisms?

Making A Microscopic Garden... A Rotting Experience

In the introductory chapter you did an informal survey of a small part of the readily visible part of your backyard. Now let's check out the invisible guests you entertain every day without being aware of them.

These creatures are given the collective name microorganisms. This is really a mixed-bag term that involves organisms from nearly all the living kingdoms, although bacteria and fungi are particularly well represented. Biologists group them together because they can use the same methods and techniques to study them.

A Typical Leeuwenhoek "Microscope"

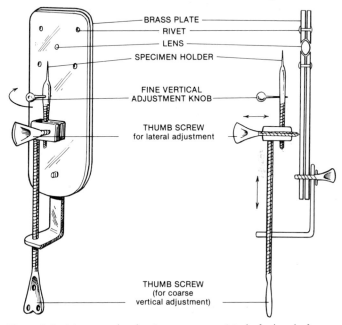

BRASS PLATE
RIVET
LENS
SPECIMEN HOLDER

FINE VERTICAL ADJUSTMENT KNOB

THUMB SCREW
for lateral adjustment

THUMB SCREW
(for coarse vertical adjustment)

Figure 3-2: A Leeuwenhoek microscope consisted of a bead of glass, finely polished, that was fitted between two brass or silver plates. The specimen was fitted on a pin in front of the lens and adjusted with thumb screws. Magnifications of several hundred power were possible. With oblique lighting techniques, creatures as small as bacteria were visible. (Drawing based on photographs and drawings by Dobell, Dover Ed., 1960)

The trick in observing such small creatures is to get them to reproduce with such success that they become visible. This will be a chance to exercise your abilities in observation and description. We'll create ten microbial gardens. You'll

Microorganisms

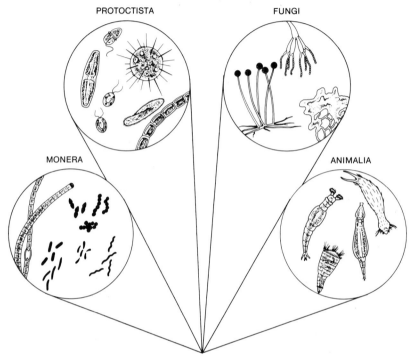

PROTOCTISTA

FUNGI

MONERA

ANIMALIA

Figure 3-3: Microorganisms have representatives in most, if not all, the living kingdoms. Monera, Protoctista, and Fungi, however, account for most of the ones you will find. Rotifers, small multicellular animals, are sometimes mistaken for protozoans.

need an observation sheet for each one. Make a heading which looks something like the example at the top of page 33.

For those without access to a microscope, delete the third column. To fill in the I.D., or identity, column you will need the use of some good keys or field guides. See the list at the end of the chapter. Some broad tips for putting organisms in major groups will be given shortly.

Some things to consider in the macroscopic or naked eye appearance are: color, size, texture, and growth form. Some appropriate adjectives might be fuzzy, cottony, powdery, smooth, rough, shiny, glistening, dull, compact, spreading, and/or irregular. Because microorganisms multiply quickly, it's also important to keep track of the times and dates of observations.

Bowl No. _____ Material in Bowl _____			
DATE	MACROSCOPIC APPEARANCE	MICROSCOPIC APPEARANCE	I.D.

Here's a technique for cultivating your garden of microorganisms that works very nicely:

Get ten containers of the same size. If you don't have glass bowls, margarine containers work fine, and you can just throw them away afterwards. This experiment is smelly, and you may want to set it up in an out-of-the-way spot. "Plant" your gardens as follows:

Bowl #1: Fruit, cut to fit in the bowl, if necessary.

Bowl #2: Slightly crushed grapes with sufficient water to cover.

Bowl # 3: Water from a lake, pond or river, containing surface and bottom material.

Bowl #4: Enough hay to cover the bottom of the bowl and 200 ml of water.

Bowl #5: A few dried beans and 200 ml of water.

Bowl #6: Cottage cheese or cream cheese, spread over the bottom of the bowl.

Bowl #7: Half or quarter of lettuce head in a little water.

Bowl #8: Two pieces of stale bread and enough water to moisten the material. The bread should not be soaked. You can cover after exposing to the air for 24 hours.

Bowl #9: Mix 5 g of cornstarch with 95 g of rich soil. While mixing the soil and starch, add enough water to give the mixture a doughy consistency. Put it in the bowl and keep the mixture moist throughout the experiment.

Bowl #10: 1 g of peppercorn and 200 ml of tap water.

All bowls should be covered, but not tightly. A plastic food wrap works well with a few fork holes poked in the top. Keep them out of direct sunlight. Bowl #1 should be isolated somewhat from the other bowls as spores produced there might contaminate the rest of the garden.

Record what happens in each of the bowls over a seven to ten day period. Look for texture, color, smell, and growth characteristics of the microorganisms that develop.

As I've said, all the species that appear in your gardens won't be "simple" organisms. However, you will find colonies of bacteria in some bowls. Bacteria represent a very early form of organism organization that lived in undisputed mastery of the planet for at least 2.5 billion years. Bacterial cells are examples of prokaryotic cells (see Chapter 2). They are significantly different from the cells of virtually every creature you're familiar with, lacking complex membranes, organelles, and some biochemical pathways that characterize eukaryotic cells. We'll take a look at this difference in more detail later, because it is one of life's fundamental innovations. For now, let's concentrate on surveying the variety to be found in your gardens by previewing some of the things you may see on a bowl-by-bowl basis.

The creatures that germinate in your gardens will fall into three basic categories which correspond pretty well with the three living kingdoms of Monera, Protoctista (also called Protista by some authorities), and Fungi. If you find a rotifer or two, the Animal kingdom will also be represented. Bacteria belong to the kingdom Monera. Their colonies can be recognized by their texture and growth pattern. They appear as small, shiny circles that may eventually coalesce as your garden ripens. Each circle started out as a single bacterium that stuck to the organic material in the bowl and found substances it could "eat." To be able to eat, or metabolize something, the bacterium must have certain enzymes (organic catalysts) that promote the breakdown of a complex substance into simpler things. During the process, energy is released. Most common bacteria, as well as other microorganisms, feed on things like sugars, starches, and cellulose. They break them down into alcohols and carbon dioxide. About every 20 minutes a bacterial cell divides, in another 20 each of those divide. The result is a pile of cells that is visible to the naked eye. Yet each individual cell is no more than one to ten millionths of a meter long.

Fungal growth, on the other hand, is like the spreading growth of tree roots. Thread-like strands called hyphae tunnel into organic material and send off shoots in all directions. When they have absorbed enough nutrients, they send up aerial strands that appear velvety or cottony in a mass. At the tips, black or colored sporangia develop that ultimately release clouds of spores into the air.

Protozoans and algae are both subgroups within the Protoctista. Protozoans are not readily evident without using a microscope, but algal growth is green and often shiny, with a rather slimy look. Rotifers are in the same size range as protozoans and can be mistaken for them at first glance. Refer to some of the guides at the end of this chapter for more detailed information.

Previewing the Garden

The rotting fruit in bowl number one may not be as unpleasant as some of the rest of your garden. Depending on which microorganisms invade it, you will get a certain amount of fermentation occurring, which yields a sweetly pungent odor. Molds often attack first, especially the cottony sort. After they liquify the fruit a bit, bacteria may become more evident. Look for sporangia on the mold as the culture ages.

You may find both bacteria and molds in dish number two. The water may turn cloudy with bacterial growth, while fungi crown the "islands" of grape fragments.

The pond water, since it starts out with its own array of protozoans, algae, crustaceans, worms, and a host of other things, may not change as dramatically as the rest.

However, the change in temperature, lighting, and oxygen content from the pond to your house probably will result in a shift in the balance between organisms. If you have a microscope, note the kinds and quantity of different organisms over the time span of your observations.

The hay-water mixture is a classic growth medium for bacteria and protozoans that feed on them. Some molds may also appear.

You should see good mold growth on the beans. Green molds like *Aspergillus* or *Penicillium* are common. The fluid part of the mixture turns cloudy with bacterial growth.

Green molds often attack cottage cheese and cream cheese. Certain molds, of course, are encouraged to colonize and "age" cheeses to enhance their flavors.

Bacteria go after lettuce readily and seem to melt it away as they multiply.

The common bread mold, *Rhizopus nigricans*, starts out as a white growth and later becomes tipped in black. Under a dissecting scope the black sporangia appear to sit like beads of obsidian on their weaving stalks. *Aspergillus*, the green mold, is also common on bread. You may see *Neurospora*, a pink-tipped mold that scientists use in genetic research.

Soils vary greatly in their ecology and so will your results, but a host of microorganisms live in healthy soil and some will use the cornstarch.

Did you have any trouble finding peppercorns? They are the dried berries of black pepper. They used to be so common that another meaning of peppercorn is something "trifling or insignificant." They are included in your gardens for "historical interest" because Leeuwenhoek found mixtures of peppercorns and water that had sat around for awhile to be a rich source of microorganisms. Leeuwenhoek describes one series of observations this way:

> "Having made sundry efforts, from time to time, to discover, if 'twere possible, the cause of the hotness or power whereby pepper affects the tongue...I did now place anew about one third ounce of whole pepper in water, and set it in my closet, with no other design than to soften the pepper, that I could the better study it. This pepper having lain about three weeks in the water, and on two several occasions snow-water having been added thereto, because the water had evaporated away; by chance observing this water on the 24th April, 1676, I saw therein, with great wonder, incredibly many very little animalcules, of divers sorts; and among others, some that were 3 or 4 times as long as broad, though their whole thickness was not, in my judgement, much thicker than one of the hairs where-with the body of a louse is beset. These creatures were provided with exceeding short thin legs in front of the head (although I can make out no head, I call this the head for the reason that it always went in front during motion). This supposed head looked as if 'twas cut off aslant, in such fashion as if a line were drawn athwart through two parallel lines, so as to make two angles, the one of 110 degrees, the other of 70 degrees. Close against the hinder end of the body lay a bright pellet, and behind this I judged the hindmost part of all was slightly cleft. These animalcules are very odd in their motions, oft-times tumbling all around sideways: and when I left the water run off them, they turned themselves as round as a top, and at the beginning of this motion changed their body into an oval, and then, when the round motion ceased, back again into their former length."

Clifford Dobell, a protozoologist who published this passage from Leeuwenhoek's letters (see references), believes that the creature Leeuwenhoek saw was *Bodo caudatus*, a small protozoan. Leeuwenhoek went on to describe additional protozoa, such as *Cyclidium* and *Vorticella*, and other organisms that were probably bacteria. What you find in your peppercorn infusions will largely depend on the source of water used, but don't be surprised if Leeuwenhoek's friends make an appearance.

Robert Hooke became Secretary of the Royal Society in 1677 and was Leeuwenhoek's contact at the time he discovered peppercorn animals. Hooke is best known for his "best seller" *Micrographica*, a book with many drawings of the microscopic world based on Hooke's own research. In basic biology texts he is mentioned as the discoverer of the cellular nature of living things. Hooke's drawing of cork cells is a classic. According to Brian J. Ford, who has recently made new discoveries from Leeuwenhoek's original letters, *Micrographica* may well have been the initial inspiration for Leeuwenhoek's life work. Hooke, however, went on to other studies, while Leeuwenhoek perfected the single lens microscope (an instrument Hooke found too tedious to use) and continued to peek into the world of the very small.

Hooke repeated Leeuwenhoek's work with peppercorns and also made infusions with barley, oats, peas, and other grains. In a letter to Leeuwenhoek, he said: "Of this the President & all the members present were satisfied & it seems very wonderful that there should be such an infinite number of animals in soe imperceptible a quantity of matter." Leeuwenhoek was informed in a later letter that even His majesty King Charles II was quite impressed with his animalcules.

During his lifetime Leeuwenhoek was to receive other honors from royalty. He was visited by King James II, Czar Peter the Great, King Frederick I of Prussia, Augustus II, and Queen Mary. It is fitting in many ways that men and women of note took an interest in Leeuwenhoek's little creatures, which would in more modern times be

Peppercorn Animals

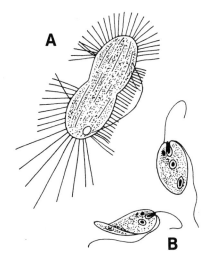

Figure 3-4: *Clifford Dobell believes that Leeuwenhoek saw species of* Cyclidium (**A**), *a ciliated protozoan, and* Bodo caudatus (**B**), *a protozoan with two flagella, in his peppercorn infusions. They are shown here drawn to the same relative scale.* Cyclidium *is about 40 microns long and* Bodo *is 11 to 22 microns long and 5 microns thick (about as thick as the hair of a louse, according to Leeuwenhoek). (Drawings based on Kudo, 5th edition)*

recognized as having taken the basic steps to colonize Earth some three and a half billion years ago.

Now that you've had a chance to see (and smell) a few of these creatures yourself, let's look at some of their accomplishments.

The Basic Innovations

The organisms that you found in your microscopic gardens had to solve some of life's most basic problems. The fact that they are around today, some four and a half billion years later, is testimony to their success. Other organisms embellished and added to their successes, but the basic chemical pathways they pioneered are fundamental to the life processes of all "higher" organisms.

The earliest bacteria and their kin arose on an Earth much different than the world you see now. In fact, the present state of the Earth is largely the result of the chemical changes that living things initiated. These chemical reactions are the result of life's need for energy, energy that can temporarily create the order of complex organic compounds out of the random distribution of inorganic elements.

The Earth was a cooling rock for millions of years after it condensed from the sun's leftover raw materials, with water vapor and noxious gases comprising its atmosphere and all the oxygen locked in chemical union with other compounds. Even under these conditions, however, it seems that life probably arose relatively quickly—say 200 million years after a hard crust formed. The building blocks of life, such things as amino acids, alcohols, and nucleic acids, form easily when mixtures of water vapor, nitrogen, methane, ammonia, carbon dioxide, and hydrogen are exposed to a source of energy, such as an electric discharge.

Ultraviolet (UV) light was abundant at the Earth's beginning. Although now it is harmful to living things, initially it was largely responsible for generating many chemical reactions that formed a wide variety of organic compounds. UV light also smashed into water molecules in the air and in the process created free hydrogen and oxygen gases. Enough oxygen ultimately was created in this way to form a layer of ozone in the upper atmosphere. The ozone layer is opaque to UV light, so gradually a partial barrier developed against this high energy radiation. The result was that larger and more complex chains of molecules could be built without rapidly breaking down to simpler substances.

Figures 3-5A & 5B: *Bacteria most like those that may have prospered on the primitive earth can be found in special habitats like the geysers and mudholes found in Yellowstone Park. They survive boiling temperatures and utilize the minerals as sources of energy.*

The work of Sydney Fox at the University of Miami has shown that under the conditions prevailing on a virgin Earth, proteins bound to clay surfaces will form spheres, which he calls proteinoids. Structurally they are amazingly similar to simple, bacterial-type cells. They are membrane-bound sacks filled with the medium they surround, which in a primitive, oxygen-free ocean would have consisted of many of life's organic precursors. Proteinoids even divide when they get too large, much like the simple fission reproduction of bacteria. It is still a large step from proteinoids to bacteria, but given the vast stretches of time available, it doesn't take a great deal of imagination to envision the transformation.

Self-replication, the key to the perpetuation of life, requires energy to assemble complex chemicals. Initially, life could have cannibalized itself when organic compounds were everywhere, courtesy of the wash of UV radiation. But when the ozone layer formed, that source of energy was reduced. Life was forced to tap other sources. The options were: use other forms of solar energy not blocked by the ozone layer, such as visible sunlight, or use inorganic compounds that are capable of "donating" hydrogen atoms to fuel reactions that could create "food" (sugars or alcohols).

In the course of time, living things explored all the alternatives open to them within the constraints of their biochemistry. This is nowhere more evident than among microscopic organisms. Some microorganisms make their own food. These are called autotrophs. Others consume each other, autotrophs, and/or their decay products. These latter organisms are heterotrophs. But among heterotrophs there are those that need oxygen to metabolically burn their food and those that use less efficient fermenting processes. Among autotrophs, some use light in several variations of photosynthesis, and others use inorganic chemicals such as nitrogen, manganese, and sulfides in their energy relations. Each of these subgroups, in turn, has variations.

If this seems confusing, it is. What can we make of it?

All bacteria and their kin, which are members of the kingdom Monera, share the same type of cellular organization. They are all considered to be prokaryotes. All other kingdoms have a eukaryotic cellular organization. The differences are profound and are a good part of the reason why living things began to change much more quickly a billion to two billion years ago (see Figure 4-9).

One of the differences between the two is that prokaryotes have enormous variations in their metabolic machinery.

Eukaryotes do not have these variations, which is what we might expect if eukaryotes developed from a particular line of prokaryotes. The four "advanced" kingdoms of organisms all share the same kind of metabolic factory for burning food. All higher autotrophs, such as green plants, use chlorophyll as a light-trapping agent in photosynthesis.

Prokaryotes are small organisms (1 to 10 micrometers). A micrometer is one millionth of a meter or one thousandth of a millimeter. Eukaryote cells are 10 to 100 micrometers long. Although a few are microscopic, most are large organisms.

DNA, the hereditary material, is not confined within membranes in prokaryotes, nor is it associated with chromosomes. Eukaryotes have their DNA bound to chromosomes within a well-defined nucleus.

Cell division is a simple process of fission in prokaryotic organisms. There are no systems to insure an equal parceling of cell contents. Sexual systems are rare and, when present, result in fragmentary transfers of genetic information. Eukaryotes divide by mitosis, a process orchestrated with cell organelles such as centrioles (in animals), mitotic spindle, and microtubules to ensure equal division of DNA to daughter cells. Sex is common, with equal participation by male and female partners. Generations alternate between one with a double chromosome number (diploid) and one with a single set of chromosomes (haploid). Fertilization creates diploids; meiosis, a type of reduction division, creates haploids.

Multicellular forms are rare in prokaryotes, and there is no association of cells into tissues. Multicellular organisms with extensive tissue development are characteristic of eukaryotes.

Many prokaryotes are killed by contact with oxygen. Most eukaryotes need oxygen to live. The ones that don't are clearly secondary modifications on the main theme.

In eukaryotes the enzymes needed to oxidize or burn food materials are isolated and organized in cell organelles called mitochondria. Prokaryotes have no mitochondria or other organelles. Their enzymes are bound at various sites on the cell membrane.

Prokaryotic cells move by the action of whip-like structures called flagella. They are composed of a single kind of protein. Flagella in eukaryotes are complex structures of tubulin and other proteins that have a characteristic "9 + 2" arrangement. There are nine pairs of microtubules arranged in a circular pattern around two single microtubules. This arrangement is found in all eukaryotes in structures as diverse as sperm tails and lung cilia.

In photosynthetic prokaryotes the enzymes for photosynthesis are bound to the cell membrane, not packaged separately. There are various kinds of photosynthesis, some depending on oxygen and some not. Various end-products may form, such as sulfur, sulfate, and oxygen. In common green plants and other eukaryotes that use photosynthesis, enzymes are packaged in membrane-bound organelles called plastids. The photosynthetic pathways are all similar and produce oxygen as a by-product. Thus, in the process of discovering self-replication and various forms of autotrophism, the microscopic members of the kingdom Monera changed the chemical makeup of the atmosphere. They also laid the groundwork for additional cellular innovations.

The development of eukaryotic cellular organization, perhaps from some sort of symbiotic association between two or more types of prokaryotes, led to much more rapid change and an even more extensive impact on our planet. Bacteria carried out their crucial tasks in anonymity before Leeuwenhoek performed his careful observations. Fortunately, we can look forward to many more such surprises. "By diligent labor," Leeuwenhoek said, "one discovereth matters that before one hath deemed inscrutable."

REFERENCES

Asimov, Isaac. 1976. *Asimov's Biographical Encyclopedia of Science and Technology.* New York: Avon Books.

BSCS Green Version. 1963. Chicago: Rand McNally and Co. The microscopic garden exercise in this chapter is from this high school biology text. The BSCS books are quite good, emphasizing an investigative approach to science.

Bullock, William. 1979. *The History of Bacteriology.* New York: Dover Publications. (Originally published in 1938 by Oxford University Press)

Dobell, Clifford. 1960. *Antony van Leeuwenhoek and His Little Animals.* New York: Dover Publications. (Originally published by John Bale, Sons and Danielsson, Ltd. in 1932) This book is considered a classic in the field, although Dobell was most interested in Leeuwenhoek's work with microorganisms. Much of Leeuwenhoek's pioneering work in other fields, such as reproductive biology, cytology, and parasitology, is not readily available in English translation.

Ford, Brian J. 1985. *Single Lens, The Story of the Simple Microscope.* New York: Harper and Row. This book approaches Leeuwenhoek largely from the viewpoint of instrumentation. Ford found unopened original specimens from Leeuwenhoek attached to his original letters that were overlooked because many later researchers were working with microfilm copies. Under the scanning electron microscope, some of Leeuwenhoek's own blood cells were found. Ford's book also provides interesting historical highlights about other early microscopists.

Lewin, Roger. 1982. *Thread of Life: The Smithsonian Looks at Evolution.* Washington, D.C.: Smithsonian Books.

Margulis, Lynn and Karlene V. Schwartz. 1982. *Five Kingdoms: An Illustrated Guide to the Phyla of Life on Earth.* San Francisco: W.H. Freeman and Co.

Schierbeek, A. 1959. *Measuring the Invisible World.* New York: Abelard-Schuman.

FIELD GUIDES

Boney, A.D. 1975. *Phytoplankton.* London: Edward Arnold Ltd.

Durrell, Gerald. 1983. *The Amateur Naturalist.* New York: Alfred A. Knopf. A worthwhile reference for any naturalist to have on his or her shelf.

Headstrom, Richard. 1983. *Adventures With Freshwater Animals.* New York: Dover Publications.

Jahn, T.L. 1979. *How to Know the Protozoa,* 2nd ed. Dubuque, Iowa: Wm. Brown Company. The "How to Know" books published by Wm. Brown Company are probably the best general purpose field guides.

Klots, Elsie B. 1966. *The New Field Book of Freshwater Life.* New York: G.P. Putnam Sons.

Kudo, Richard R. 1977. *Protozoology,* 5th ed. Springfield, Illinois: Charles C. Thomas. A more comprehensive, technical text.

Prescott, G.W. 1978. *How to Know the Fresh-Water Algae,* 3rd. ed. Dubuque, Iowa: Wm. Brown Company.

Smith, Gilbert M. 1950. *The Fresh-Water Algae of the United States.* New York: McGraw Hill Book Company.

4

Exploring Inner Space

Everyone knew approximately what to expect when astronauts first saw the Earth from a distant vantage point in space: a blue and white marbled sphere hanging in obsidian blackness. Nevertheless, the first photographs inspired awe and reflection. Our perspective of our planet, and ourselves, was forever altered.

As we have seen, in 1674 a totally earthbound man, Antony van Leeuwenhoek, experienced a similar change in perspective. He looked through a carefully ground lens of his own manufacture and discovered a world of minute "animalcules" whose existence was never suspected. The creatures flourished in staggering numbers and a bewildering variety of forms. Leeuwenhoek found his tiny animals and plants everywhere—in the water, the soil, the air, and in, on, and around the human body.

probably need a microscope to pursue travels to these worlds, but for now, thinking small requires only imagination—a cheap, but very effective vehicle to other perceptions.

If you're six feet tall, for example, think of yourself shrunk to one tenth of that size, or 7.2 inches. You are now in a position to consider pole vaulting with a pencil. If you were ten times smaller still, you would be a thimble-sized three-quarters of an inch high. A small jumping spider would be the size of your terrier (and much more dangerous). Another power of ten reduction and you're nearly out of sight at seven hundredths of an inch. A paper clip, lying flat on the desk, would be more than waist high. One more factor of ten—you are now ten thousand times smaller than when you started—gets you into the world of inner space. Here are a few of the sights you'll see in this microscopic world:

Thinking Small

The microscopic wonders of "inner space" can generate a consuming fascination—especially if you can make the perspective change complete by "thinking small." So, before I give you some suggestions for setting out on your own explorations, let's preview some of the creatures you may see. Eventually you will

Vorticella *Forest*

Vorticella forests are easy to locate (see Figure 4-1). Search them out on the side of a fish tank or in a quiet pond. They take up far less room than a national forest of equal diversity. *Vorticella* are stalked animals that secure themselves to a network of algae in some tranquil spot and draw in their dinner with a whirlpool of

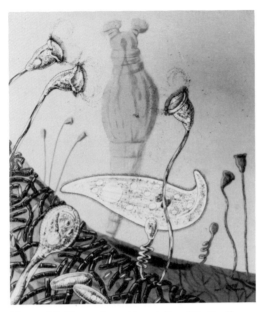

Figure 4-1: *The spring-action stalks of* Vorticella *contract into a coil when the organism is threatened by a predator, in this case a rotifer lurking in the background.*

Figure 4-2: *The long spines of* Actinosphaerium *paralyze unsuspecting visitors. A large freshwater alga, appropriately called the water net, serves as the backdrop for this microscopic drama.*

water created by the rapid beating of hair-like cilia around their "mouths." Their stalks, which act very much like springs, contract into a coil with a bulb at one end when *Vorticella* is threatened or disturbed. A variety of small animals can be found drifting and darting among the stalks, looking for even smaller protozoans, bacteria, and diatoms to lunch on.

Stellate Predators

Some predators in the micro world take life even easier than *Vorticella*. *Actinosphaerium*, with its long, delicate, needle-like spines, lies in wait or drifts like a land mine among ropes of green algae (see Figure 4-2). When small animals or plants run into its spines they are paralyzed and drawn toward the spherical body by the conveyor-like action of cytoplasm surrounding the spines. At the end of their trip the victims are engulfed and digested at leisure.

Ceratium Spaceships

Ceratium is an interesting creature whose relatively large size, projecting

Figure 4-3: A long flagellum propels Ceratium as it cruises through its watery world. Despite its animal-like appearance, it contains the pigments and cellulose characteristic of plant cells.

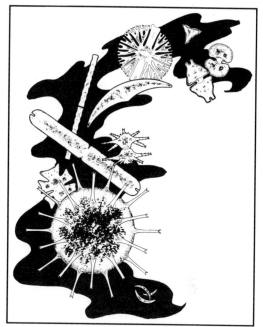

Figure 4-4: Most desmids reproduce asexually by splitting at the isthmus that separates their mirror-image semi-cells. Sexual reproduction results in zygotes—spined spheres that eventually become new individuals.

horns, and armor-plated shell give it the appearance of a spaceship cruising the local cluster (see Figure 4-3). *Ceratium* and its cousins were once considered animals because of the long whip-like flagella that move them through the water and the red, light-sensitive eyespot that many of them have. These creatures, however, sport golden brown pigments for making sugar from sunlight and their armor is made of cellulose—the stuff of plant cells.

Some forms are stalked, clinging to filaments of algae, but the majority are free-living, solitary cells. Most have a groove around their midsection housing a flagellum that gives them a spinning motion. A longer, trailing flagellum propels them forward. The total effect is that of a colorful corkscrew boring through the water.

Bouquets of Desmids

Desmids are diverse and beautiful algae (see Figure 4-4). Most share the characteristic, however, of being divided into two mirror image halves, or semi-cells, divided by a narrow isthmus. Desmids also store their food as droplets of starch which often appear as smooth beads in the cytoplasm. Many species appear yellowish in color because of iron compounds they accumulate in their cell walls. Sex in desmids results in the formation of beautiful, spined spheres (zygotes) that later "hatch" into new individuals. Reproduction without sex is more common.

They split at the isthmus and each half grows a new semi-cell.

Although most desmids are solitary, a few species form delicate unbranched filaments. In silhouette, they could almost be the backbones of some aquatic serpent. In *Micrasterias* (see Figure 4-6), the cells overlap and lock together by means of tiny hooks on the large central lobes, looking much like butterflies linked by dark, double arrowheads. Iron compounds fringe the cells with yellow.

Closterium cells are brilliant jade cres-

Figure 4-5: Some of the most elegant desmids belong to the genus Micrasterias. *The lobes of these cells are deeply indented and their edges are intricately carved, much like snowflakes.*

Figure 4-6: The chain of cells in the foreground with one individual detached is a species of Micrasterias. *In the upper left,* Staurastrum *in three-quarter view appears to be a pair of spiny pyramids connected at their tips. The same desmid (lower right) appears quite different from front and end views.*

Figures 4-7 *(illustration)* **& 4-8** *(photo): Beads of starch dot the length of a Closterium cell. Small granules of gypsum are housed at either end of this crescent-shaped desmid.*

cents when seen against a dark background (see Figure 4-7). They are desmids, although the individual cells are not composed of the usual mirror image halves. Glistening beads of starch line up along a furrow in the green chloroplast. At either end of the cell, small granules of gypsum, perhaps having some function in cell orientation, dance with brownian motion. *Closterium* reproduces by breaking in two at the center of the cell. Two cells may separate while close to each other and their contents fuse in a sexual union to form a zygote.

Some of the creatures you have previewed are producers, analogous to the more complex and easily seen green plants we see every day. They make their own food from raw materials and sunlight. In terms of function they are not unlike some kinds of photosynthetic bacteria and cyanobacteria. Others are consumers, analogous to fungi and animals like ourselves. These two options were explored very early in the history of life on Earth, as we shall see shortly, and served to displace bacterial types of life forms from their cosmopolitan dominance of our planet to much more restricted niches.

To Be or Not
To Be Solar-Powered

It's difficult for us to imagine the world as it must have been some three billion years ago when bacteria were the highest form of life on Earth. It was a world unlivable and barren by our standards, with little or no free oxygen and no land life at all—unless you count hot springs and volcanoes where chemotrophic bacteria could survive on types of photosynthesis dependent on hydrogen sulfide or other minerals. In the oceans, similar kinds of photosynthetic bacteria thrived in huge colonies that grew into pillar-shaped forms called stromatolites. Stromatolites are formed when the mucus-like secretions of bacteria trap sediments. As the cells become covered they migrate up toward the light, rising on a self-made column of sand and mud.

Perhaps one of the first men to witness a scene very much like those ancient times was the English explorer and pirate William Dampier. In 1699 he set out on a voyage up the western coast of Australia. About five hundred miles north of the modern town of Perth, he sailed into a large bay. The numerous sharks in the waters impressed him and he named the site Shark Bay, but he almost certainly also saw in the shallow lagoons the large, cushion-like stromatolites rising from the water.

The stromatolites that grow in Shark Bay and places like southern Florida and the Bahamas are not exactly like those ancient ones now found only as fossils. The configuration is nearly identical, but modern stromatolites are formed by the growth of Cyanobacteria (blue-green algae) in environments that are so salt-rich that other organisms can't survive there. Cyanobacteria, like more advanced green plants, carry on photosynthesis with the help of the green pigment chlorophyll rather than the reddish brown pigment desulfoviridin. Both processes allow organisms to take carbon dioxide from the air and make sugars and starches for growth and reproduction. But photosynthesis accomplished with the help of chlorophyll yields oxygen as a by-product. Without oxygen no animal life as we know it could have developed.

Chlorophyll may have become the catalyst of choice for photosynthesis before complex eukaryotic cells developed. All known plants, as well as some prokaryotic creatures like the Cyanobacteria, use the same biochemical pathways for photosynthesis. If eukaryotic cells developed, as some scientists believe, from the symbiotic association of two or more pro-karyotic cells, one of the latter must have possessed the biochemical pathway to the photosynthesis we see today. The exact time that this occurred, of course, can not be determined, but we know approximately when it happened be-cause the Earth rusted.

Mars is a planet that is rusted now. When you know where to look, you can see its red glow in the night sky. Mars has oxyen, but most is tied up in the oxides of iron and other elements. When the same highly reactive element was first produced on Earth as a by-product of photosyn-thesis, the same thing occurred. Because most bacteria lived in their sediment high-rises in the oceans, the oxygen combined with soluble iron there (ferrous iron) to form the insoluble ferric iron. Ferric iron precipitated out to cover the ocean bot-toms. Today we recognize this material as the banded iron formations. They have an age of slightly more than two billion years.

Eukaryotic Inventions:
Sex and Predation

As important as chlorophyll-mediated photosynthesis was in the transformation of Earth, it wasn't the innovation that started life on the "fast-paced" living that characterizes organisms today. Stromato-lites were the monuments of early life for billions of undisturbed years. And they might still dominate the Earth today if it weren't for the development of preda-tion. The stromatolites exist in Shark Bay because there are insufficient predators to graze them away. According to Steven Stanley, a paleontologist at Johns Hop-kins, predation opens up ecological job opportunities for other kinds of creatures. The steady-state situation created in a "filled-up" environment becomes impos-sible.

One example of a basic eukaryotic grazer is a little protozoan called *Tetra-hymena*. Although I first met this bacteria-eater in college, he and his relatives have been in somebody's backyard for a billion and a half years or so. *Tetrahymena* is a pear-shaped organism with longitudinal rows of hair-like cilia that make him look a little like a moving pin cushion. He's com-monly found in fresh water, but has also been known to take up residence in the innards of the garden slug and, oc-casionally, people. We'll discuss how to collect *Tetrahymena* shortly, but first I should also introduce the tiny oval or-ganism called *Chlamydomonas*, credited by some scientists as the inventor of sex.

Chlamydomonas is one of the plant-like members of the Protoctist kingdom. It, too, is a somewhat pear-shaped cell, but it

Comparing Common Microorganisms

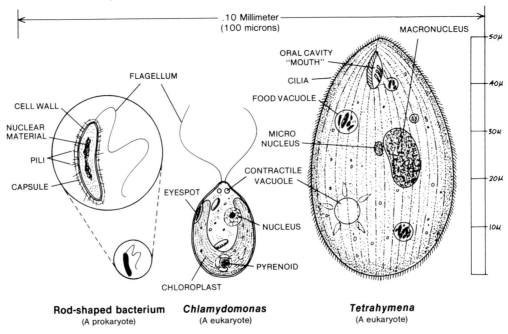

Figure 4-9: *These three creatures represent a typical range of sizes for microorganisms. Note that the eukaryotic cells are vastly more complex than prokaryotic ones, with "organelles" that may be the remnants of prokaryotic parasites that became dependent on their hosts. Chloroplasts and mitochondria, in particular, have their own set of genes that replicate independently of those in the nucleus.*

is surrounded by a thin cellulose wall and makes its own food with the help of a cup-shaped chloroplast not unlike those of higher plants. Instead of rows of cilia, like *Tetrahymena*, it has two whip-like flagella at the front end that are as long or longer than its body. *Chlamydomonas* sports a bright red eye spot, however, that can be found in various parts of the cell, depending on species. *Chlamydomonas* can reproduce asexually by dividing longitudinally. Daughter cells remain within the "mother's" cell wall until they develop

walls of their own and the "wiring circuits" for their eyes and other organelles.

Sex is resorted to when *Chlamydomonas* is hungry. The specific cue is a deficiency of nitrogen. Although *Chlamydomonas* doesn't have distinct sexes, as such, there are + and − mating types. These mating types have different versions of the protein agglutinin on their flagella. The different agglutinin molecules stick to each other when the flagella of different strains meet. This event seems to cue the cells to dissolve their cellulose walls. A

Chlamydomonas: A Sexual Pioneer

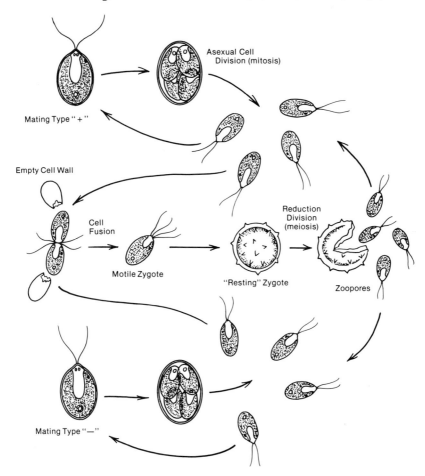

Figure 4-10: Chlamydomonas *can reproduce sexually or asexually. Sexual reproduction mixes the genetic information from two strains, producing individuals that may have a survival advantage under variable environmental conditions. (Life cycle based on drawing in Grobstein)*

protuberance swells from the surface of one cell and initiates fusion—a very elemental example of fertilization. Division then takes place, but it is the special kind of division called meiosis: a process where chromosome-borne genes mingle and intertwine in a double division that halves the number of parent chromosomes.

Why Sex?

"Why not?" you might say. Sex is an expensive process. Expensive in terms of the energy organisms spend to prepare for it and perform it. Many organisms quite literally die for it. Others make themselves vulnerable by sporting bright colors to catch the interest of a mate or by courtship routines that expose them to predators. Why pay such a price? The reason lies with the intertwining chromosomes we find in meiosis.

As we've already seen, primitive life arose when it became possible to reproduce complex chemical systems in minute detail. The helically wound DNA molecule can split in half longitudinally, and each half serves as the template to replace the other half. Bacteria utilized this basic machinery to great effect in colonizing the Earth and filling it to capacity. Changes, or mutations, could occur to slowly effect alterations in living cells, but on the whole, self replication is and was a very conservative and accurate process. Sex is a process that shuffles the genetic cards. Two organisms fuse their genetic heritages during fertilization, mix them up during the crossing over and intertwining of chromosomes that then takes place, and in a series of divisions, reduce the total amount of genetic information to what it was in one individual.

The new organism that results is not a twin of its parents, but a combination of both. When considering an entire population of organisms, sex produces many more variations than simple cell division can. Under optimal conditions these many variations might be at a disadvantage to the cells that had fine-tuned their genetic messages over eons. But in the normal, shifting conditions of the Earth, a major change in climate could totally wipe out a population of organisms that had identical characteristics. A sexually produced population, however, might contain variants that were tolerant to the new conditions.

The Beginning of Cellular Cooperation

Chlamydomonas and its relatives also display some basic experiments with another important innovation: multicellularity. *Chlamydomonas* lives its life as a solitary cell (although daughter cells occasionally may fail to separate and will form groups of four or more cells). In the same family, however, an organism called *Pandorina* exists normally in colonies of sixteen. Each of its individual cells could easily be mistaken for *Chlamydomonas* if seen singly. Nevertheless, in *Pandorina* there is already evidence of sexual special-

Volvox and Its Relatives

(Drawn to the same scale)

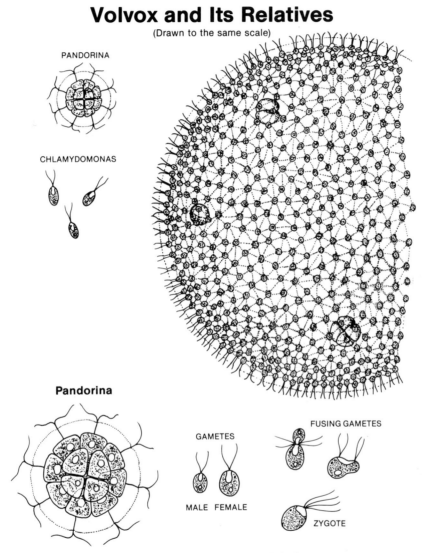

PANDORINA

CHLAMYDOMONAS

Pandorina

GAMETES

MALE FEMALE

FUSING GAMETES

ZYGOTE

Figure 4-11: *The* Volvocales *family may reflect some of the living "experiments" on the way to multicellularity. These are some of the most beautiful green algae to see under relatively low magnification.* Volvox *is about the size of a grain of sand.*

ization. One gamete is significantly larger than the other, although both have flagella and are free swimming.

Another creature in the same family is called *Eudorina* and is similar to *Pandorina* except that there is even more sexual differentiation. The large gamete, which might now be called an "egg cell," is non-motile. The smaller gamete, or "sperm cell," is small and flagellated. This family of organisms, the *Volvocales*, gets its name from an organism called *Volvox*, which is a spherical assembly of 500 to 40,000 *Chlamydomonas*-like cells interlinked by threads of cytoplasm. In *Volvox*, only certain cells of the colony can give rise to egg cells and certain others to sperm cells. The latter cells are released by one colony and must penetrate another to fuse with an egg cell.

The animal-like members of the Protoctist Kingdom may have had other routes to multicellularity. You'll notice in Figure 4-9, which is drawn to scale, that *Tetrahymena* is a considerably bigger organism than *Chlamydomonas*, even though both are single, eukaryotic cells. *Tetrahymena* and other complex ciliated protozoans also have a more complicated "nervous system" to coordinate their fast-moving predatory habits. Perhaps partly for this reason (and partly because of a more intricate sex life) they have more than one nucleus. They have a large macronucleus and one or more micronuclei.

When an organism has many nuclei but no cell walls dividing them, it is called a syncytium. As we shall see later, fungi have made a successful life style out of this sort of organization, but it is rare among animals. Mesozoans, a small group of animals that parasitize the octopus and other invertebrates, spend part of their lives as syncytial organisms—large cells with a scattering of nuclear "control centers." These creatures, lacking any organs or organ systems, range in size from about .5 mm (the size of a large *Volvox* colony) to 8 mm (about a third of an inch).

Whether a syncytial organization or some sort of colonial assemblage of cells led to multicellularity is a question that may never be answered—although one should use the word "never" sparingly. There are even simpler creatures than the Mesozoans, for example, about which little is known. A species called *Trichoplax adhaerens* looks very much like a ciliate, with its surface covered by cilia. It is made up of two layers of cells, however, a few thousand in number. It lacks nearly everything else you associate with animals, including tissues, organs, organ systems, a head and a tail end, a left and a right side, and even a mouth. *Trichoplax* is a pancake of cells that glides about in the ocean on a suitable surface, perhaps indenting itself from time to time to make a temporary stomach cavity into which it pours enzymes and from which it absorbs a meal.

Although you won't find *Trichoplax* in your backyard you may find another simple multicelled animal when you col-

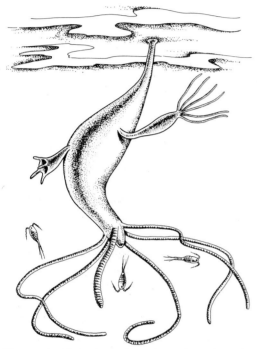

Figure 4-12: Hydra *hangs from the undersurface of a film of water and stretches its tentacles to grab a passing meal. Stinging nematocysts can paralyze a victim so that the tentacles can haul it in. A young hydra is "budding off" from one side of its parent.*

lect *Tetrahymena* and his cousins: the hydra. *Hydra* has a mouth, three layers of cells, and some interesting offensive equipment in the form of poisoned "darts" called nematocysts. *Hydra* has no anus, however, and what can't be digested in its simple sack-like stomach must exit from the mouth. When relaxed,

Hydra is long and slender, rather pinkish in color, and hangs beneath the surface of calm water with wispy tentacles hanging down, ready to entrap an unwary meal.

And now it's time for us to do some hunting!

Stalking Wild Protoctistans

To find algae and protozoans you can arm yourself with something as simple as a glass jar or go all out for grappling hooks and plankton nets. I will outline most of the options and discuss some of the problems you may encounter.

Equipment

Microscope: You'll need a microscope to see most protoctistans. Microscopes are fairly expensive, costing about as much as a good 35mm camera, but they are your ticket to a completely new world. If your major interests lie with bigger plants and animals, you may want to invest in a dissecting microscope rather than a compound microscope. The latter type usually magnifies from 100 to 1,000 times whereas the former enlarges things from 10 to 40 times. At forty power magnification you can't see the very small forms or the detail of larger protozoans, but you do see ciliates like *Tetrahymena* and *Paramecium* as three-dimensional creatures diving and spinning in their small world. Compound microscopes give you greater detail, but at the expense of viewing the organism in relation to other creatures and surrounding objects.

Probably the most common place to buy microscopes is a biological supply house (see Chapter 1), but there are a few other sources. School districts sometimes auction off outdated equipment to the public. The best bet would be to make a contact with a science teacher or check with the administration building for auction dates. Check prices on new equipment beforehand so you'll recognize a reasonable price. Companies that repair microscopes often take old scopes in on exchange, and you may be able to make a purchase from one of them. If you live in a university town you might contact someone in a biology department who could let you know of equipment being replaced and for sale. Government surplus auctions also occasionally feature microscopes.

Even with a hand lens or "linen tester" with a magnification of 10 times, however, you can see a great deal and shouldn't forego a few aquatic field trips.

A Drag: A drag is a grapnel hook of sorts used for snagging submerged water plants that serve as the home for many micro creatures. You can make one by fashioning coat hanger wire into a three-pronged hook and tying it together with lighter wire. Attach a long cord and perhaps a lead "sinker" used in fishing. You can then spin the hook around your head and fling it out into lakes or streams.

Dredge Nets: Dredge nets are made of tough nylon attached to a hose pipe frame that can withstand dragging through mud to pick up samples of bottom-dwelling creatures.

Dip and Scoop Nets: Scoop nets are used in fishing to land "the big one," and dip nets can be bought at pet stores. They are useful for scooping up floating weeds or sampling fine sediment.

Plankton Nets: Plankton nets are used to sample the surface waters of lakes. They are available from biological supply houses or they can be made by using silk bolting cloth net (No. 20 mesh) from an importer or flour mill. A brass or stout wire ring, 6 to 8 inches in diameter, should be used for the mouth of the net. Cut the silk so that when attached to the ring a cone about 14 inches long is formed. Do not attach the silk directly to the ring, but sew it to a piece of muslin which is, in turn, sewn over the ring. Either use the net as a closed cone or remove the tip of the cone about ½ inch or less from the end and insert a small vial (4–6 dram capacity),

Some Useful Equipment

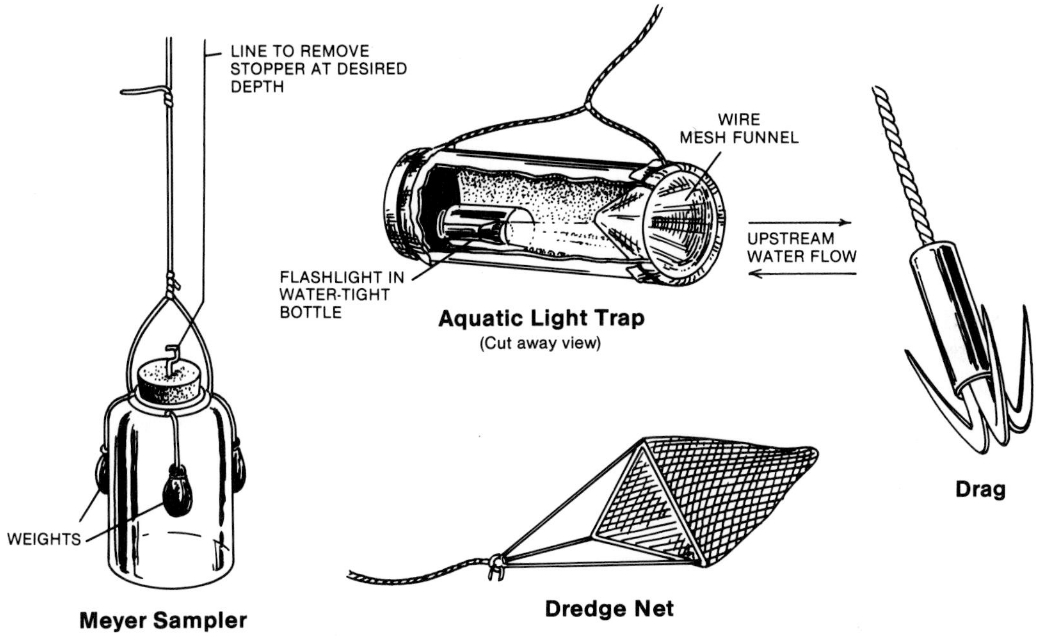

LINE TO REMOVE
STOPPER AT DESIRED
DEPTH

WIRE
MESH FUNNEL

FLASHLIGHT IN
WATER-TIGHT
BOTTLE

UPSTREAM
WATER FLOW

Aquatic Light Trap
(Cut away view)

Drag

WEIGHTS

Meyer Sampler

Dredge Net

Figure 4-13: *The equipment illustrated here may be useful in aquatic collecting. Materials can be home-made or ordered from supply houses. (Based on drawings in Durrell)*

Plankton Net

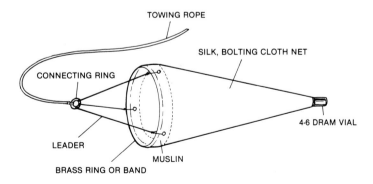

TOWING ROPE

SILK, BOLTING CLOTH NET

CONNECTING RING

4-6 DRAM VIAL

LEADER

MUSLIN

BRASS RING OR BAND

Figure 4-14: *Plankton nets concentrate surface species of organisms that would be hard to find otherwise.*

which can be tied about its neck into the apex of the net.

If you use the simpler closed cone, you must turn the net inside out and rinse the contents into a bottle or dish. If you have the vial at the apex of the net, you just need to reverse the net and pour the concentrated contents of the vial into a collecting bottle.

Aquatic Light Traps: Aquatic light traps can be used to sample stream or water organisms that are drawn toward light. Take an old flashlight and seal it in a watertight jar. Put the jar in a section of earthen pipe. Cover one end and create a funnel for the other out of a fine mesh wire net. Attach the assembly to a strong cord and lower it into the water facing upstream.

Meyer Sampler: A Meyer sampler allows you to sample deeper lake water. Attach a strong cord to a jar that can be fitted with a cork. Attach a second line to the cork. The bottle will have to be fitted with weights so that it will sink when full of air. When the jar reaches the sampling depth, pull the cork out. If you pull the jar up slowly, not much of the deep-water sample will be displaced.

Glass Jars: Glass jars are really the mainstay of aquatic collecting. Plastic is also all right as long as the containers haven't housed caustic chemicals. Metal containers should be avoided, for microorganisms are often sensitive to metal ions.

Where to Go

Fresh-water sources in your backyard are most likely to be temporary pools of standing water. Our friend *Chlamydomonas* can be found there, as well as protozoans and the large colonial *Volvox*. In fact, Leeuwenhoek first discovered *Volvox* in water from "ditches and runnels . . . and on coming home, while I was busy looking at the multifarious very little animalcules a-swimming in this water, I saw floating in it, and seeming to move of themselves, a great many green round particles, of the bigness of sand-grains." However, fresh water is also found in ponds, lakes, springs, streams, and other flowing water, and even trapped within or on plants.

Standing water is often covered with dense mats of algae. A green filamentous alga, called *Spirogyra* because of the spirally, twisted nature of their choloroplasts, is a common type. *Spirogyra* appears as "clouds" of cottony growth that can be scooped up with a jar or net. Many animals and other plants like the glass-shelled diatoms live with or on *Spirogyra*, so when you collect it you really get a whole community of life. Blue-green algae such as *Nostoc* are also common. *Nostoc* is filamentous, but the colonies grow in gelatinous olive drab balls easily seen with the naked eye. Diatoms may form a brownish layer on the surface of

Common Water Plants

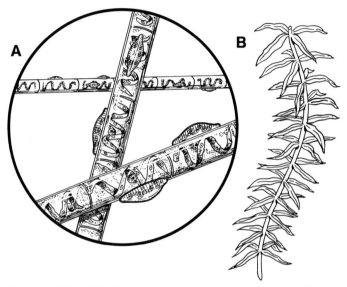

Figures 4-15A & 15B: Diatoms and other algae are often found attached to other water plants. In quiet, fresh water they may live on the surface of the filamentous green alga, Spyrogyra (**A**). **B** is an illustration of Elodea, a common plant found in streams, that also may harbor an assortment of microscopic creatures.

sand grains.

The **flowing water** of rivers and streams has its own assemblage of creatures. You can use your drag line to hook some kinds of algae and submerged grasses. *Elodea* is a common flowing-water species that is popular for aquaria. Various algae such as diatoms live on their stems. Other kinds of algae cling to rocks and dams. To get representatives of them you may have to place a stone or brick in front of the sampling site to divert the flow of water while you scrape some into a jar. Another trick is to string a fine mesh screen downstream in the water and then have someone start turning over rocks upstream. You will also get plenty of insect larvae and other invertebrates this way.

Lakes are best sampled with the plankton net and the Meyer bottle. Plankton is the name given to the free-floating forms of algae and protozoans that live in the upper layers of large bodies of water. Many have beautiful silicious spines and other adaptations that keep them afloat. The dinoflagellate called *Ceratium* that we

mud an inch or so below the surface of still water. They can be carefully scraped into a collecting jar with a spoon or, if you have the time, you can carefully spread a piece of white cloth over the mud. Under the influence of sunlight the motile species will migrate through the mesh. You can then collect them in a bottle. The first technique may get you other beasts, such as the single-celled *Amoeba* or its cousin *Arcella* that lives in a house of cemented

looked at earlier is especially elegant, with its three spines and a long neck that waves back and forth slowly as it twists through the water.

Dry land can be a legitimate hunting ground if you know where to look. Brickwork, rocky walls, cliffs, and tree bark may harbor diatoms, small green algae, and some kinds of blue-greens (cyanobacteria). Sphagnum and other mosses, as well as lichens, often have microscopic house guests. Also check flowering plants like pitcher plants, which have reservoirs that collect rain water.

Some Notes on Culturing

Successful culturing requires mostly common sense and a little knowledge of what microorganisms need to thrive. Remember pHOTWaFL and all should go well. pHOTWaFL stands for pH, oxygen, temperature, water medium, food, and light.

pH is a measure of the acidity of a liquid (see box). The lower the pH number the more acidic a solution is. A pH of 7 is considered a neutral pH. Most organisms like a pH value near 7. The majority of water samples you collect will be in this range. The acidity level can change, however, due to an accumulation of decay toxins, evaporation of the water level, or contamination by something in your collection containers. pH will usually take care of itself if you follow the other suggestions below.

Most common microorganisms need **oxygen** to survive, just as you do. Therefore, keeping cultures in dishes that are rather broad relative to depth is best to maximize the surface area in contact with air. Since this also maximizes evaporation, you have to keep adding fresh pond or distilled water, and/or place loose-fitting sheets of glass over the dishes. Supply houses make culture dishes of varying diameters that are stackable, so that you only have to cover the top one. Flowing-water organisms need to be kept in an aerated container of some kind. The "bubblers" used in fish aquaria will serve the purpose.

Most microorganisms, since they can't regulate their own internal **temperature**, will die at temperatures greater than 85°F. Therefore, don't place cultures in direct sunlight or in rooms that get very hot. Optimum temperatures lie between 60–70°F. Although temperatures between 70 and 85°F are not lethal, they cause a general increase in metabolism that in turn leads to more pressures on oxygen and food supplies and to faster build-up of waste products. Warm water also holds less dissolved oxygen than cold water. Some creatures can "deactivate" themselves when conditions are bad by encysting, but others cannot.

The **water medium** is critical to microorganisms. Use distilled water, spring water, or pond water. Chlorinated tap water can be fatal to microorganisms. Use glass, plastic, or earthenware containers, because metal containers may "leak" toxic ions. Wash containers with soap and water, then rinse them thoroughly,

pH

To denote the acidity or alkalinity of a liquid, scientists use an expression called pH. A liquid which is exactly neutral, neither acidic nor alkaline, has a pH of 7. Acid waters have pH values below 7 and alkaline waters have pH values above 7. The scale below will provide some common referents.

ph			
		7	Pure water
0		8	Sea water
			Alkaline soil
1	Stomach fluids	9	
	Volcanic soils		Alkaline lakes
2	Lemon juice	10	
	Vinegar		Soap solutions
3		11	Household ammonia
	Tomatoes		
4		12	Saturated solution of lime
5		13	
6		14	

making the last rinse with distilled water. Don't use containers that have previously held toxic materials such as cleaning agents or preservatives.

Food becomes most critical if you're trying to make pure cultures of protozoans. Most protozoans are either bacteria eaters or consumers of other protozoans. One source of bacteria is the decaying stems, leaves, or seeds of plants. (Be careful—too rich a source of decay is bad for some species.) Malt tablets, bread crumbs, or rice grains can also be used in sparing quantities to support such protozoans as our friend *Tetrahymena* or the ever-popular *Paramecium*.

Green protoctistans can pretty much support themselves with an indirect **light** source, such as a north window. A few kinds, like *Euglena*, are "switch hitters" and will lose their chloroplasts and take up bacteria munching if circumstances dictate. Protozoans, too, are best kept in an indirect light source as many of them feed on their green brethren.

Perhaps one of the most enjoyable things to do is to fill a large jar with pond or lake water, making sure there is at least some algae included, and then watch the succession of animal and plant species you find over several months or even a year. From the darting specks you can see with the naked eye to the tiniest ciliates spinning away beneath the microscope, you will be delightfully surprised.

REFERENCES

Attenborough, David. 1979. *Life on Earth.* Boston: Little, Brown and Co.

Durrell, Gerald. 1983. *A Practical Guide for the Amateur Naturalist.* New York: Alfred A. Knopf.

Grobstein, Clifford. 1965. *The Strategy of Life.* San Francisco: W.H. Freeman and Co. A nice little volume that discusses general biological principles and contains good illustrations.

Lewin, Roger. 1982. *Thread of Life: The Smithsonian Looks at Evolution.* Washington, D.C.: Smithsonian Books.

Margulis, Lynn and Dorion Sagan. 1986. *Microcosmos.* New York: Summit Books, a division of Simon and Schuster.

Morholt, Brandwein, Joseph. 1966. *A Sourcebook for the Biological Sciences,* 2nd ed. New York: Harcourt Brace Javanovich. An excellent reference for the naturalist. Most biology teachers who got their training twenty years ago are probably familiar with it. It is still available from Harcourt Brace Javanovich.

Morse, Gardener. 1984. 3-D Algae Sex. *Science News,* vol. 126, no. 6 (August 11).

Raham, R. Gary. 1979. Stalking the Wild Diatom. *The American Biology Teacher,* vol. 41, no. 8 (November). For those interested in diatoms, my article discusses diatom hunting and biology in more detail.

Wolle, Francis. 1884. *Desmids of the United States.* Bethlehem, Pennsylvania: Moravian Publication Office. This book may be a hundred years old, but it has beautiful hand-colored pictures of algae. If you live near a college library, try to look it up. Wolle also did books on other algal groups.

5

Killer Toadstools and Their Kin

The Toadstool Mystique

Toadstools have a certain fairyland image about them. I suppose there is nothing particularly wrong about this, although it is misleading. Even the name "toadstool" conjures up pictures of a frog prince sitting on a diminutive chair waiting for a princess to happen by. Fairy rings—circles of bright green grass associated with bands of mushrooms at certain times and seasons—were long considered to be the dancing grounds of fairies and elves. The "toad" in toadstool, however, is more likely a derivation of the German word Töd, which means death.

The word "toadstool," when used in the United States, usually refers to an inedible mushroom. In Great Britain the only fungus that qualifies as a mushroom is the commercially cultivated species you find in supermarkets. There are many species of large, edible fungus, however, with the "toadstool" shape that many people find quite tasty. The price for misidentification, nevertheless, is high. Many fungal toxins are quite deadly. Common names for several species of *Amanita* like "death cap," "fool's mushroom," and "destroying angel" provide a clue to their characteristics. Some *Amanitas* taste rather good, I understand, and others can produce hallucinogenic sensations, and so people find reasons for taking risks.

Mushrooms and toadstools make up a large class of organisms in one of the three phyla of fungi, the Basidiomycetes. All members of this phylum bear their spores on club-like cells called basidia. Like other fungi, toadstools cannot produce their own food as green plants do and thus are consumers. In this regard they are like animals. Because they are small and are not free to move around, however, de-

Figure 5-1: Amanita muscaria, *also called the fly agaric, is found under conifers and birch. It is shown here at various stages of development. The cap is bright red; the scales, stalk, spores and warts are white (the warts are occasionally yellow). The volva is scaly with concentric rings.*

The Kingdom of Fungi

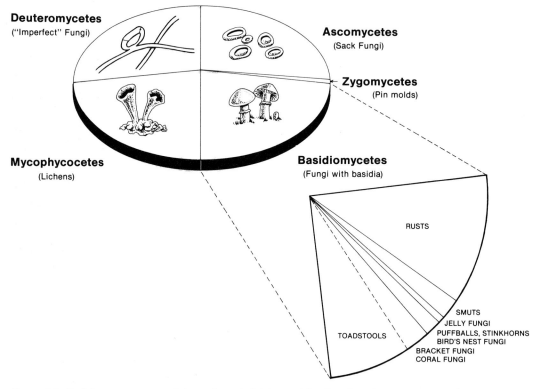

Figure 5-2: *Basidiomycetes, one of the major subdivisions of the Kingdom Fungi, comprise about one fourth of all species. Approximately a third of Basidiomycetes are toadstools and bracket fungi. Deuteromycetes are something of a group of "leftovers." Species are placed here when no sexual phases are discovered that would put them in another group!*

composition of dead and dying organisms is their main source of energy rather than predation or grazing. Nevertheless, toadstools and their relatives have recently been found to actively trap and consume animals on occasion—an unsuspected turn of events that further helps to shatter their benign image. More on this later. Toadstool ecology is, in fact, quite complex. Some species are parasites of plants or animals; other species form symbiotic associations with trees in unions called mycorrhizae; and as we'll see in the next chapter, some form essentially permanent associations with algae and are called lichens.

The toadstool itself is merely one phase in the life of organisms that are very important in the cycling of elements in nature. Moreover, toadstool biology is fascinating and unique—a fact which is partially reflected in the elevation of fungi, in recent years, to "Kingdom" status in the scheme of classification. So before you aim your lawnmower in their direction and shred them out of the bluegrass, pick up a toadstool and take a closer look.

The Toadstool Body Plan

The fungus which eventually produces a toadstool spends most of its lifetime as fine white threads called hyphae that wend their way through the interstices of the soil or invade organic material such as tree stumps and feed on what they can find there. Hyphae are so thin that they are invisible without magnification, but seen in large masses they are cottony in appearance. Collectively, they are referred to as the fungal mycelium. Only when temperature, humidity, and other factors are right does the mycelium produce the toadstool structure, or sporophore—literally, an organ which bears spores. And that is the toadstool's only function: to bear and disperse billions of spores so that a few will find a favorable place to germinate and start another feeding mycelium.

It's virtually impossible to tell one fungus from another by looking at a mycelium. Toadstools are recognized by

Figure 5-3A & 3B: Bracket fungi, like the one growing on this log, (A) and puffballs, (B) so named because of the cloud of spores that puff out when you poke them, are non-mushroom types of Basidiomycetes.

The Toadstool Body Plan

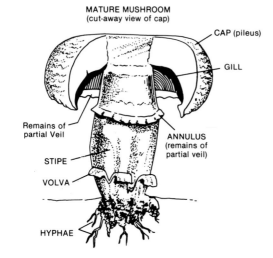

MATURE MUSHROOM
(cut-away view of cap)

CAP (pileus)

GILL

Remains of partial Veil

ANNULUS
(remains of partial veil)

STIPE

VOLVA

HYPHAE

Toadstool Growth

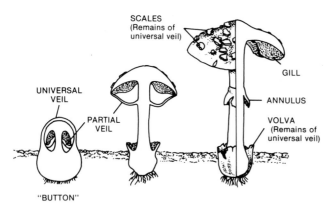

SCALES
(Remains of universal veil)

GILL

UNIVERSAL VEIL

PARTIAL VEIL

ANNULUS

VOLVA
(Remains of universal veil)

"BUTTON"

Figure 5-4: *The illustration of the toadstool body plan is based largely on A. bisporus, although the cultivated mushroom does not have a volva. The total spore-bearing part of the gill surface is called the hymenium.*

The diagram of toadstool growth demonstrates how the tissues of the young "button" are torn to create features in the adult toadstool. All toadstools will not display all features. (The diagram is redrawn from Ingold, 1979)

their sporophore characteristics, which fortunately are fairly simple. A toadstool starts out as a small "button." The button is entirely covered by a sheet of tissue called the universal veil. When the toadstool grows, this veil will tear, sometimes leaving "scraps" on the cap (pileus) called scales, and fragments at the base of the stem (stipe) called the volva. The fly agaric *(Amanita muscaria),* which is common under conifers and birch, shows these characters well. Its cap is red or orange and the scales stand out white against the bright color. Although this *Amanita,* by the way, isn't the most deadly of its species, it can be poisonous. A person's individual body chemistry, as well as poorly understood variations in the chemistry of this fungus, lead to effects ranging from hallucinations and euphoria to nausea and death. Experimentation is not recommended. Its common name comes from a European custom of soaking the cap in milk to kill or stupefy flies.

As you may recall, spores are

produced on special cells called basidia. The basidia line the surfaces of the gills that radiate outward from the stipe to the outer margins of the cap. In some toadstools (the Boletes) the gills are replaced by pores and the basidia line the tubes to which these pores lead. The basidia-bearing structures on yet other toadstools have a toothlike appearance. The spore-bearing surface, whatever it may look like, is referred to as the hymenium. The young gills in a growing toadstool are covered by a sheet of tissue called the partial veil. When this tears it often leaves a ring, or annulus, around the stipe. This can also be easily seen in various species of *Amanita*. The base of a toadstool is often very important in identification because its shape as well as the appearance of the volva is distinctive. When collecting for identification, take care to get the whole toadstool and not just cap and stipe.

Toadstools and the L.T.U.

Toadstools have a reputation for quick growth, and this contention can be supported by the backyard biologist. We all know that toadstools appear in less than one L.T.U. (the Lawnmowing Time Unit). The length of this time unit varies, but usually is seven days. One week the lawn will be free of toadstools, and the next they will be scattered all over. In fact, one gets the impression that toadstools appear virtually overnight. This is almost the truth. Under ideal conditions it takes about five days for a toadstool to grow from a "button" to its adult form, but most of that growth occurs during the last six to nine hours when the cells of the stipe elongate dramatically. There is much to be learned about the specific environmental cues for toadstool growth and the hormonal controls that orchestrate the cooperative growth of hyphae into the complex mushroom form.

The mushroom form, however, is admirably adapted to its job of dispersing spores. The cap protects the delicate gills from direct bombardment by rain drops while still creating a micro-environment high in humidity so that the tissues don't dry out. The stipe raises the cap far enough above the earth so that the spores can be caught by eddies of air and carried to more distant places. Moreover, gill structure and growth of the stipe are coordinated to maximize the number of spores that eventually reach the ground in a place favorable for growth. To understand this coordination of structure and growth (while keeping in mind that human motives can't really be applied to any other living thing), think of yourself as a

Gill Structure

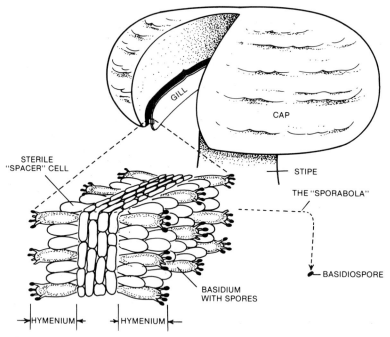

Figure 5-5: *The spore-bearing cells of toadstools called basidia line both sides of each gill. The gills are spaced so that a spore fired from a basidium flies out about halfway to the next gill and then falls straight down. The right-angled path traversed by the spore is sometimes referred to as a "sporabola."*

spore looking for a good place to grow. You have one goal: to reach the ground to begin growing and feeding in the rich loam. You are perched on a fragile stalk that connects to a large, club-shaped cell beneath you. That cell, the basidium, is in turn attached, along with many others like itself, to a brownish wall that seems to stretch to infinity in either direction. Far across from you is another wall, a neighboring gill that is packed with other basidia and their complement of four spores each. Below you is a vast chasm that

extends down to a misty sea of grass. Suddenly, a blister forms at the base of one of your three sister spores. It swells rapidly and bursts. Your sister disappears momentarily, only to be seen once again near the middle of the chasm between gills. Then she begins falling—a slow motion descent of about half a centimeter a second that carries her down and away. The light from below makes her one glittering speck among many. A blister forms beneath you. It swells and breaks, tearing your stalk and casting you violently into the emptiness. If you went too far you would hit the opposite gill and stick uselessly to another spore or a basidium. You only go out about half way, however, and then begin falling. If the gills were not exactly parallel, you might hit either yours or the neighboring one farther down and again be stuck before you even get out of the cap. But you fall freely and, as you leave the dark cavernous cap, are swirled away in a whirlpool of air.

The growing toadstool is sensitive to several environmental cues that keep the gills oriented properly with respect to each other and the ground. The young, growing stipe is sensitive to light and moves toward it. This allows a toadstool germinating in a tree stump, for example, the maximum opportunity to create a cap out in the open where spores are more readily dispersed. As the toadstool matures it loses its sensitivity to light and becomes negatively geotropic—it grows away from the direction of the pull of gravity. The gills, in turn, grow toward the pull of gravity and, in addition, are usually tapered so that they are narrower at the bottom. All these factors work together to help insure that the descending spore clears the cap when it falls.

You can get a feeling for the melodrama we've enacted if you take a toadstool and cut off the stem close to the cap. Pin the cap, gill surface up, to a thin piece of cork.

Invert the cork over a jar or beaker and turn out the room light. Shine a tight beam of light through the beaker and you should soon see a cascade of twinkling specks— each speck a tiny basidiospore. A field mushroom with a cap 8cm across produces an estimated 600,000 spores per minute over a period of perhaps four days. The ink cap fungus, found commonly on dung as well as on lawns, and named because its margins rot away as mature spores are released, was estimated to produce 5,240,000,000 spores in 48 hours or 1,600,000 per minute. Obviously only a tiny percentage germinates. The toadstool plays a game of success through surplus.

Spore prints are another way to demonstrate the same thing and are often used to help identify species. Again, remove the

Figure 5-6: Spore prints are easy to make, attractive, and useful in identifying toadstools (see text).

stipe as close to the cap as possible and set the cap, gill side down, on a piece of white paper. Cover with a bowl to minimize drafts. In a couple of hours there will have been enough spores discharged to create patterns of radiating lines corresponding to the regions of most dense spore fall. Since the bottom edges of gills have no basidia, the area below them will be blank. Dark bands of spores will outline these regions with a dusting of spores between adjacent bands from basidia in the tissue connecting the gills. The spores may be white, cream, buff, brown, pink, purple, or black. If you splice light and dark pieces of paper together and then place the cap so that it bridges the junction, the spores will show up better on one side or the other, depending on their color. To preserve a spore print, use a can of spray adhesive like 3M's Scotch brand Spray Mount (available at art supply stores), and spray above the print so that the adhesive drifts down onto the paper without scattering the spores. You can then lay a thin sheet of clear mylar over the print. Alternatively, you could just spray the print with an artist's varnish in the same manner.

To get a notion of how the spore is "fired off" from the basidium, inflate a balloon and set it on top of an open jar. Inflate a second, smaller balloon and hold it next to the first with a ringstand. With sharp scissors, cut the tied base of the small balloon and the sudden escape of air will propel the first off the top of the jar. It's believed the blister at the base of a basidispore is filled with a gas, but there are some competing theories. Other fungi have techniques that involve liquid under pressure, the sudden rounding off of a turgid spore, and even electrostatic repulsion. At any rate, the spores are fired off in prodigious numbers, because the conditions for their germination into active hyphae must be just right when they reach the ground.

To Grow or Not to Grow . . .

As you may have guessed, water is very crucial for toadstool growth. A well-watered lawn is much more likely to yield a crop of fungus. Wind can be particularly harmful once toadstools have formed because it rapidly robs them of their moisture reserves. Toadstools also need a source of organic carbon, of course, in the form of sugars, starches, or cellulose, and this is available from grass clippings, leaves, dead insects, and other materials readily at hand. A source of nitrogen is necessary, usually in the form of amino acids that are liberated when proteins decay. Some fungi can use inorganic sources such as nitrates or ammonium salts. Sulphur is gleaned from sulphates, and phosphorus from phosphates, both courtesy of soil bacteria. Potassium and magnesium are needed in milligram amounts (thousandths of a gram), whereas trace elements like iron, zinc, copper, and possibly manganese and molybdenum, are needed in microgram quantities (millionths of a gram). Unlike green plants, toadstools don't need a source of calcium. Most fungi can make most of the vitamins they need, but some toadstools require B_1, biotin, thiamin and pyridoxin.

Much of what scientists know about toadstool nutrition comes from working with the ink cap fungus, *Coprinus cinereus*. Here is a "brew" that has been used successfully for laboratory culturing:

```
glucose ......................................... 1.0 g
dl-a-alanine (an amino acid) ........0.1 g
dipotassium phosphate ..............0.2 g
magnesium sulphate ................0.02 g
thiamin hydrochloride ....... 50 micro g
water .......................................... 100 g
```

But a proper diet isn't everything. pH is one additional factor important for growth. The laboratory medium above comes out just slightly alkaline, which is good for *C. cinereus*, but most toadstools like things slightly acidic.

Temperature turns out to be a crucial factor. Minimum ranges for growth are about 2–3°C, optimum ranges between 20° and 30°C, and the maximum temperatures tolerated are between 35° and 40°C. You'll notice that there is a fairly broad range from minimum to optimum, but a narrow range between optimum and too hot. These temperatures reflect optima for growth, however, and not necessarily optima for sporophore (toadstool) formation. In the honey fungus, *Armillaria mellea*, for example, the temperature has to be down around 15°C before the typical fruiting body appears.

Light is a somewhat unexpected requirement for toadstool production in a number of species. The ink cap, for example, needs at least a short exposure to the blue end of the spectrum. The common cultivated mushroom, *Agaricus bisporus*, however, has no such requirement. If you wish to watch mushroom growth, by the way, and get something nice to eat after your studies, you can buy mushroom farms of *A. bisporus* from outlets like Carolina Biological Supply Co. You receive a box with specially prepared soil containing the mushroom mycelium to which you must add a "casing material," mostly peat moss, which Carolina provides. Follow the supplier's directions for temperature and watering regimes and you get mushrooms within a month. Several crops can subsequently be harvested.

The nutrients which the mycelium absorbs are moved throughout the entire fungus very efficiently, even though there are no specialized cells for translocation as in green plants. The toadstool mycelium is basically a single, interconnected body of protoplasm. Although the mycelium appears to be a simple filamentous network of cells, the crosswalls are pierced by pores, called dolipores, through which protoplasm and even cell nuclei can move with ease. This arrangement is a fundamental departure from both plant and animal strategies. Fungi also make use of chitin, the same hard material usually associated with insect skeletons, whereas green plants do not. In toadstools, water and nutrients appear to move up the center of the stipe and fan out radially once in the cap. With the necessary raw materials at hand, a toadstool can fulfill its reproductive potential.

The Sex Life of Toadstools

If you take one basidiospore of a typical toadstool, such as the ink cap fungus, and place it on a suitable growth medium, it will germinate and produce a mycelium. You will never get a toadstool, however. A basidiospore has one set of genetic information, just as an egg cell or sperm cell does in mammals like ourselves. The basidiospore is thus haploid and the mycelium which it produces is a monokaryon. This word literally means "one nut" in the Greek from which it is derived—the "nut," in this case, being a haploid nucleus. To make a toadstool, two compatible monokaryons must meet and fuse to form a dikaryon—a mycelium that eventually has two haploid nuclei in each of its cells. In higher animals and plants these nuclei might be considered analagous to egg and sperm cells, but unlike egg and sperm, these nuclei will coexist side by side for a long time without merging to form a zygote. The nuclei will, in fact, divide simultaneously but independently, creating new mycelium and distributing themselves at least through the outer fringes of the original monokaryons. This association of nuclei of different genetic composition in the creation of the fungal "body" is called heterokaryosis and is characteristic of fungi in general.

You can't see nuclei in the unstained strands of hyphae that make up the mycelium, but you can often tell monokaryons and dikaryons apart because of curious structures called clamp connections. Clamp connections are rather like switching tracks in a railroad system and arise for much the same reasons. Fungal hyphae are narrow enough that the

Heterokaryon Formation in the Ink Cap Fungus (*Coprinus*)

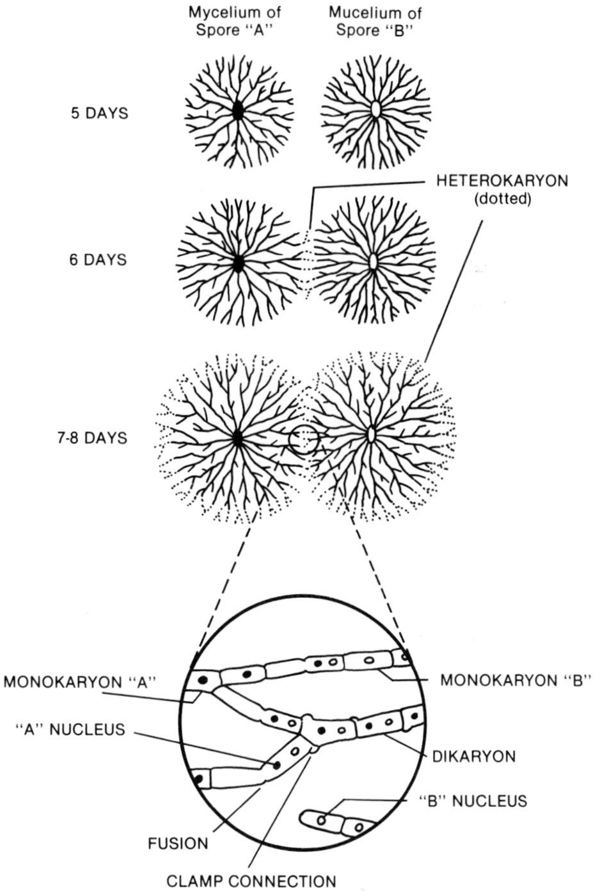

two nuclei are positioned like beads in a string rather than side by side. If both nuclei simply divided and a new crosswall formed between them, you would soon have hyphae with nuclei that are identical genetically. This defeats the advantages of heterokaryosis and doesn't occur. Instead, one of the new division nuclei at the growing end of the hyphae migrates to a bulge that occurs between it and the adjacent nucleus. One of the division products of this adjacent nucleus moves forward to join the first, non-migrating, nucleus. The bulge, meanwhile, grows and loops back until it fuses with the hypha on the opposite side of the newly-forming cell wall. The division products of the original adjacent nuclei are now assorted and segregated by the new cell. This maintains the genetic mix of the dikaryon. If this seems confusing, please consult the diagrams and things should clear up considerably.

It's still not quite clear to scientists how these nuclei move so readily from cell to cell. As I've said, there's a pore between cells, but it's fairly complex in structure and not a mere win-

Mycelium of Spore "A"   Mucelium of Spore "B"

5 DAYS

6 DAYS

HETEROKARYON (dotted)

7-8 DAYS

MONOKARYON "A"   MONOKARYON "B"

"A" NUCLEUS

DIKARYON

"B" NUCLEUS

FUSION

CLAMP CONNECTION

Figure 5-7: Single spores have one nucleus, and the hyphae that grow out radially from them are thus monokaryons. When monokaryotic hyphae from different strains meet, they fuse to form dikaryons. The dikaryotic condition spreads rather quickly to the outer margins of the combined hyphal network.

dow. In the ink cap fungus, the monokaryon grows at a rate of 2–3 mm a day at 20°C, but nuclei can migrate 20 mm a

Formation of Clamp Connections

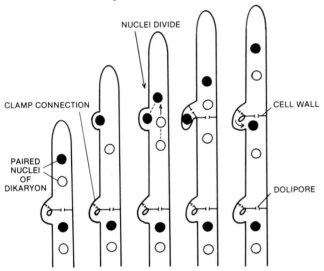

NUCLEI DIVIDE

CLAMP CONNECTION

CELL WALL

PAIRED
NUCLEI
OF
DIKARYON

DOLIPORE

Figure 5-8: *Clamp connections are the remains of a procedure that maintain the dikaryotic condition in growing hyphae (see text for explanation).*

day under the same conditions.

A dikaryon, under the physical and nutritional conditions I've described earlier, can produce a sporophore (toadstool). Hyphae, that normally grow radially away from each other, come together to form the "button" that will, in several days, expand to become a toadstool. In the spore-bearing hymenial tissue the club-shaped basidia form, and it is in these cells that the nuclei will finally fuse to form a brief diploid nucleus. Of course, all the while the nuclei have coexisted in the mycelium, their combined genetic heritage has been expressed—there just have been two "command centers" instead of one.

The diploid state lasts only long enough to carry out the reduction division of meiosis which yields four haploid nuclei. A vacuole forms and expands in the basidium while stalks sprout from its blunt tip. The four nuclei migrate to their respective stalks and develop as basidiospores. Variations exist on this pattern, of course. Sometimes the haploid nuclei reproduce once by mitosis, leaving four nuclei behind within the body of the basidium. Some basidiomycetes produce less than four spores, such as the cultivated *A. bisporus* mushroom which produces two.

Fungal genetics is a complex subject about which volumes have been written. I won't get into detail here, although the reference by Ingold will give you an introduction to toadstool genetics (and standard genetic texts usually cover other kinds of fungi like *Neurospora crassa*, the pink bread mold). Toadstools are rather unique in having their mating types determined by a pair of genes on two different chromosomes. Each of these genes in turn has variations which allow for many possible combinations. This system seems to have given toadstools great evolutionary flexibility and may help explain the num-

Formation of Mature Basidium and Basidiospores

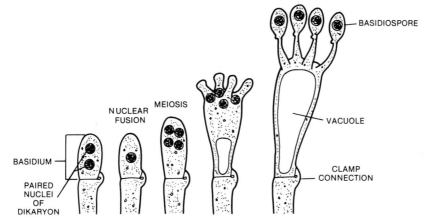

BASIDIOSPORE

NUCLEAR FUSION

MEIOSIS

VACUOLE

BASIDIUM

CLAMP CONNECTION

PAIRED NUCLEI OF DIKARYON

Figure 5-9: *The short diploid phase of a toadstool's life takes place when the nuclei of the dikaryon fuse in the young basidium. Meiosis (reduction division) yields four haploid nuclei. Usually these four nuclei each becomes a basidiospore, but in some species there may be more or less than four spores, depending on whether the nuclei undergo mitosis and how many nuclei degenerate. (Illustration based on drawings in Ingold)*

ber of different kinds that will greet you during a casual walk through the woods.

Toadstools in The Web of Nature

Before we get back to talking about the "killer toadstools" mentioned earlier, it would be good to talk about some of the more prosaic jobs that toadstools perform in nature. We all know that toadstools can be decomposers like the dung-eating ink cap fungus and many, like *Armillaria mellea*, the honey fungus, are parasitic on trees. While these are important roles,

they only represent a portion of the relationships of toadstools with other creatures. Many times the same fungus will be both parasite and symbiont, depending on which organism it's interacting with.

This latter situation is true of the honey fungus. It can be a serious pest in forests when it invades the stump of a dead tree, where its mycelium proliferates and then sends out runners to discover the living roots of nearby trees. There it invades the tree's cambium layer and can kill it in several years. *A. mellea*, however, is also one of the toadstool fungi that forms symbiotic associations called mycorrhizae with such plants as orchids. All orchids form such liaisons with fungi. Their seeds have little stored food and won't germinate without fungus partners unless simple sugars, inorganic nutrients, and vitamins are provided. In nature, the fungus apparently provides these items. The fungus actively invades orchid root tissues, but is prevented from attacking other tissues by defensive chemicals released by the orchid. The orchid then acquires some of the growth factors it needs by digesting the invading fungus hyphae. It may acquire other nutrients by absorbing excess material the fungus excretes.

A great many common toadstool fungi are engaged in a slightly different mycorrhizal relationship with trees. These mycorrhizae surround the roots of trees with a sheath of hyphae, but don't invade the root cells directly. Unlike the situation with orchids, it is the fungus that can't break down complex starches and celluloses into sugars and acquires the latter from trees. The trees receive mineral nutrients much more efficiently with their fungus coatings—particularly phosphates. *Amanita, Boletus, Tricholoma, Russula,* and *Lactarius* species are all common toadstools that are usually restricted to habitats near their host trees.

Grassland fungi are responsible for the so called "fairy rings" of dark green grass you may find occasionally in your lawns or nearby meadows. Old legends attributed their existence to the dancing activities of the "Little People" on appropriate spring evenings. Women were in danger of losing their fair complexions if they stepped in the center of one. Actually, you can think of a fairy ring as a slow motion "ripple effect," similar to the ripples from a stone dropped in clear water. What you have to drop, however, is a spore from a toadstool like *Marasmius oreades.* That spore must successfully germinate and grow. Its mycelium will grow in a radial fashion as much as 30 to 50 cm a season.

At the growing edge of the fungus, the grass will be very green, perhaps because of ammonium ions or other compounds released by *Marasmius.* Just inside the green area, however, will be a zone of dying grass. Here the fungus is either competing for nutrients to the grass's disadvantage or leaving a trail of growth-stunting waste products. When conditions are right, the toadstool stage of the fungus will appear in the "dead zone."

The center of the fungus dies off, which results in the ring shape. Partial or incomplete rings can also form. Fairy rings often are long-lived, however, and some that are 200 m across must be about 400 years old.

Marasmius oreades is a preferred food in France and Britain. It can be confused with poisonous ring-makers, however, and you should consult a good mushroom guide before sampling these "fairy footsteps."

Killer Fungi

At least one mushroom, recent studies have shown, doesn't have to rely on the stupidity or ignorance of animals to dine on fresh sources of protein. The Oyster mushroom, *Pleurotus ostreatus,* can actively paralyze and attack animals—at least the small round worms called nematodes. George L. Barron and his co-workers at the University of Guelph in Ontario, Canada, made this discovery in 1984. The Oyster mushroom is well known and is cultivated commercially in places because it is tasty. No one knew of its carnivorous habits until Dr. Barron placed some on a laboratory gel and found that its hyphae release a chemical that paralyzes nematodes within 30 seconds. The hyphae grow rapidly toward the worm and, when they reach a body orifice, enter and begin to grow. Within a day the nematode is digested. Dr. Barron speculates that nitrogen may become a limiting factor to growth on rotting wood because of scarcity or intense competition with bacteria and other microbes.

Although this is new behavior for toadstools, it really isn't unusual behavior for some other fungi. Various "simple" soil fungi have been known to trap, snare, and paralyze nematodes, rotifers, and protozoans for some time. Take, for example, the case of *Arthrobotrys anchonia,* a fungi that lassos twisting, flailing nematodes as they make their way through the soil. When nematodes are not around, these fungi have rather standard-looking hyphae. In the presence of nematodes, however, by a mechanism not well understood, the hyphae produce three-celled loops. When a worm, by chance, pokes its head or tail into one, the loop immediately constricts (within 1/10th second). The ring then develops a growth called a haustorium, which enters the worm and eats it alive. This occurs even if the ring is torn away from the parent hypha. Other species of ring fungi have "passive" rings that don't constrict, but the result is the same.

Other types of soil fungi produce sticky "pegs" that are specific for trapping worms or rotifers. Once stuck, an animal rarely escapes, but is filled with coiled threads that digest it from within. In addition to sticky pegs, *Cephaliophora navicularis* has canoe-shaped spores (called conidia) that also may produce adhesive nipples that adhere to the outer shell of rotifers and grow their way into the animal. This drama

Lassoing an Eel-worm for Supper

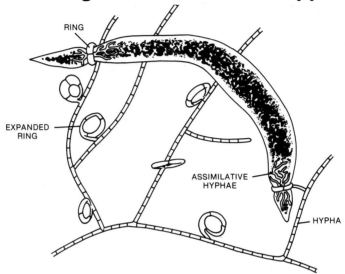

RING

EXPANDED
RING

ASSIMILATIVE
HYPHAE

HYPHA

Figure 5-10: *Soil fungi, like* Arthrobotrys anchonia, *produce three-celled loops in the presence of eel-worms. When an eel-worm blunders into one, it constricts, and hyphae grow out from the ring to digest the animal. (Drawing based on photo in Brodie, 1978)*

routinely takes place in the cow manure that can be found near some backyards.

Dr. Barron also discovered a fungus, *Haptoglossa mirabilis*, that "shoots" rotifers that brush up against special "gun cells." Cellular material shoots out under pressure and pierces the animal's cuticle and, once again, provides entry for a hungry fungal filament.

Some Notes On Collecting

Should you wish to wander beyond the bounds of your backyard to search out toadstools and mushrooms, you will be rewarded with a large and colorful variety of forms. If you want to eat some of your discoveries, however, take the time to learn these fungi well or you can get into fatal trouble. Mushroom gourmets are either knowledgeable or dead (or perhaps very lucky). Even edible species spoil rapidly, and bacterial toxins like botulism can accumulate. Other species can cause trouble only if they are not cooked properly. Alcohol and certain fungi are sometimes incompatible. Find good keys, such as those listed for this chapter, and when in doubt, eat broccoli instead.

When you have some experience in

A Rotifer-eating Fungus
(*Cephaliophora navicularis*)

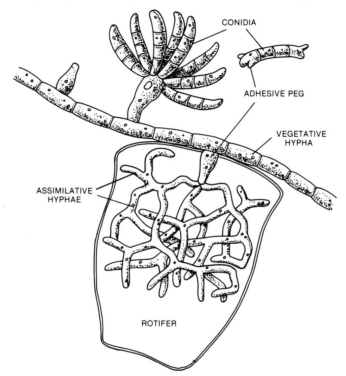

CONIDIA

ADHESIVE PEG

VEGETATIVE
HYPHA

ASSIMILATIVE
HYPHAE

ROTIFER

Figure 5-11: The soil fungus C. navicularis *produces adhesive pegs that stick to small, browsing animals like rotifers. Hyphae grow out from the pegs and digest the animal. Reproductive structures called conidia are also adhesive, and deadly, for passing animals. (Drawing based on photomicrographs in Tzean and Barron, 1983)*

observing the main characters of a toadstool necessary for identification, make yourself a checklist and note characteristics in the same order each time. Things like presence and form of volva; nature of the ring on the stipe (annulus), if any; stipe texture and shape; structure of spore-bearing tissue (gills, pores, teeth); and color, shape, and presence of scales on cap are all things to consider. Make sure you collect all the mushroom—especially the base—as the character of the volva is often very important.

Mushrooms spoil very easily once picked. Don't collect in metal or plastic containers—especially in plastic bags. Collect in wicker-type baskets, or fold the mushroom loosely in a square of wax paper and twist the ends.

Mushrooms and other fungi can be

Figure 5-12A & 12B: *Nestled among the blades of grass in your lawn you may find mushrooms of the genus* Lepiota. *They are relatives of the edible* Agaricus *mushroom, but many small* Lepiota *species are poisonous. The gills of* Lepiotas *are white rather than brown and so are their spores. They have a ring (annulus) on the stipe, but no volva at the base. Ants sometimes cultivate* Lepiotas *and you may see the mushrooms sticking out of an ant hill.*

great subjects for the nature photographer. Mostly it becomes a matter of being aware of when and what to look for. Let yourself think small and a cluster of toadstools becomes a colorful canopy that spreads over a wealth of small life that exists on the forest floor. You may wish to look up a book listed in the references called *Photographing Nature: Fungi* by Heather Angel.

Respect, Not Fear

Fungi have been part of nature since well into the Precambrian, some 570 million years ago. Their 80,000 plus living species are essential to the proper functioning of the biosphere—the shell of life about our planet. They degrade dead organisms to material that can be further decomposed by microbes; they interact with plants in very important symbiotic associations; they cause their share of diseases in a host of animals and plants; and they enter food chains by eating and being eaten in their turn. Fungi deserve our respect and wonder, but not fear.

During the course of human history, various rusts and smuts have ravaged agricultural crops from time to time. These fungi are parasitic relatives of the toadstools we've been discussing. The ancient

Romans, for example, thought rusts were diseases brought about by the wrath of the gods. They even designated one god as Robigus—the god of rust. Each year they held a festival, the Robigalia, in his honor. The Romans may have carried respect a bit too far. On the other hand, if you ever need an excuse for a party . . .

REFERENCES

Angel, Heather. 1975. *Photographing Nature: Fungi.* Hertfordshire, England: Fountain Press, Argus Books, Ltd.

Arora, David. 1979. *Mushrooms Demystified.* Berkeley, California: Ten Speed Press.

Barron, G.L. 1983. Structure and Biology of a New Harposporium-Attacking Bdelloid Rotifers. *Canadian Journal of Botany,* vol. 61.

Brodie, Harold J. 1978. *Fungi: Delight of Curiosity.* Toronto: University of Toronto Press.

Cooke, Roderic. 1977. *The Biology of Symbiotic Fungi.* New York: John Wiley and Sons.

Dickinson, Colin and John Lucas, eds. 1982. *VNR Color Dictionary of Mushrooms.* New York: Van Nostrand Reinhold Company.

Friedman, Sara Ann. 1986. *Celebrating the Wild Mushroom.* New York: Dodd, Mead & Company.

Harley, J.L. 1971. *Mycorrhiza.* London: Oxford University Press. (Oxford Biology Reader)

Ingold, C.T. 1979. *The Nature of Toadstools.* Baltimore: University Park Press.

Kibby, Geoffrey. 1979. *Mushrooms and Toadstools: A Field Guide.* New York: Oxford University Press.

Pacioni, Giovanni. 1981. *Simon and Schuster's Guide to Mushrooms.* New York: Simon and Schuster.

Science News staff. Attack of the Worm-Eating Mushrooms. *Science News,* April 7, 1984. (Original research reported in *Science,* April 6, 1984)

Science News staff. "Peacekeeper" Fungus: Rotifers Beware. *Science News,* January 8, 1983. (Original research reported in *Science,* December 17, 1982)

Tzean, S.S. and G.L. Barron. 1983. A New Predatory Hyphomycete Capturing Bdelloid Rotifers in Soil. *Canadian Journal of Botany,* vol. 61.

6

The Lichen Liaison

No one knows precisely when living things made the transition from water to land. Meager fossil remains of both plants and animals have been found in sediments that were laid down in Silurian times some 425 million years ago. By 400 million years ago land plants were established, and they shared their new frontier with arthropods, fungi, and primitive amphibians.

Plants undoubtedly preceded animals onto dry land, but perhaps not by much. If nothing else, various parasites probably "danced in step" with their hosts as the latter explored virgin territory. All the land invaders faced similar problems. They needed to find ways to avoid drying out; to solve the difficulties of gas exchange; to survive the pull of gravity without the support of a watery medium; to reproduce successfully, and disperse the new generation; and to endure much more severe swings in temperature.

The initial invasion of land could well have been a cooperative venture, not only between individual species but perhaps between members of different kingdoms. Green plants and fungi, for example, have a long history of cooperation that continues today. As we saw in Chapter 5, fungi that form toadstools have critical liaisons with trees and other "higher" plants.

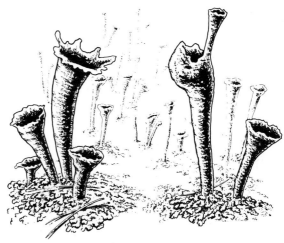

Figure 6-1: *If you were a centimeter or two tall, a patch of the lichen called* Cladonia *might look like this: a forest of pale green, misshapen trumpets that stand stiffly erect and carpet the soil. Lichens, associations of fungi and algae, may have been some of the earliest plant invaders of land.*

Another green plant-fungus association that is even more intimate is that of the lichen.

Lichens are a unique association in that the lichen plant body is totally different from that of either of its members living independently. Lichens are a combination of an alga and a fungus. Several types of algae and fungi can form lichens, and sometimes prokaryotes like bacteria may be involved in three-way unions. Algae, of course, are green plants that make sugars from carbon dioxide and sunlight. Fungi have no light-gathering pigments and

The Lichen Association

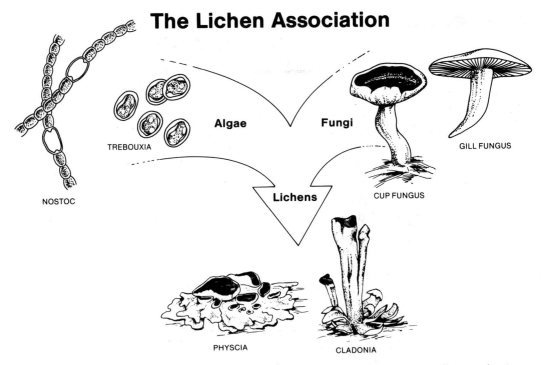

ALGAE — TREBOUXIA — NOSTOC — Fungi — GILL FUNGUS — CUP FUNGUS — Lichens — PHYSCIA — CLADONIA

Figure 6-2: *Many species of algae and fungi can form lichen associations. Trebouxia, a small green alga, is probably the most common algal partner, however, and various cup fungi (ascomycetes) are the most common fungal partner. Although the lichen plant body is unique, it is made up primarily of fungal tissue with a layer of algae near the surface.*

must feed on other plants and animals. They are very efficient, however, in extracting water and minerals from the environment.

In some ways it's not surprising that algae and fungi have "gotten together." There are some 45,000 species of free-living fungi and 21,000 known species of algae. In many places they share the same habitat, are about the same size, and face similar problems. The algae produce sugars that fungi need; the fungi "mine" the water and minerals that often limit algal growth. As independent organisms, algae and fungi might be expected to expand at the expense of each other, but many times they have united successfully until today there are some 20,000 species of lichen. Scientists are still working out the details of this co-existence.

Whether or not lichens were first to invade land, they highlight some of the problems of "terrestrinauts" and at least one set of solutions. As you may have guessed by now, some of these fascinating organisms are hiding out in your backyard.

Recognizing Lichens

Do you have a "moss rock" fireplace or stone retaining wall? If so, you have a rich source of lichens because the "mosses" on such rock are actually colonies of lichens. Do you have trees that have pale green, hair-like tufts hanging from the branches? These lichens, found where the average humidity is fairly high, are appropriately called beard lichens. Look near the ground in meadows or in clearings in forested areas. If you see soft carpets of fuzzy, green growth, you are probably looking at mosses. Often, mixed among the moss, you will see paler, "wrinkled," leaf-like plants that are colonies of foliose lichens. In fact, lichens are broken into three major groups based on the three distinct growth forms I've just described.

The foliose type look much like leaf lettuce, but their texture is more like brittle leather. They are attached to soil, rock, or bark by root-like structures called rhizines. Lichens that grow like a thin crust on rock surfaces are called crustose lichens. They are often brightly colored shades of red, yellow, and green. The color pigments may help shield the algae in the plant partnership from too much sun. The third growth form type, represented by the beard lichen, are fruticose lichens. These lichens are upright on stalks, pendent, or spread-out on their growing surface. One of the common upright types, *Cladonia cristatella*, is topped with bright red caps that are easy to spot in an open field. Other *Cladonias* have upright structures that look like tiny party horns or chalices.

Although lichens are pretty easy to recognize, there are a few simple plants with which they can be confused. I've already mentioned mosses that often grow near or even beneath some forms of foliose or fruticose lichens. However, mosses are almost always a rich green color, whereas lichens are commonly pale. Mosses grow in soft dense mats, often with thin-stalked sporangia sticking up from them like little pods on wires. Also, if you look closely at an individual moss plant, you will see a stem with a whorl of leaf-like structures around it. Lichens are always hard or leathery, often with dark, cup-like structures on the foliose types. Neither mosses nor lichens have true stems and leaves.

Some kinds of liverworts are more easily confused with lichens, but they're not nearly as common. Liverworts are semi-aquatic plants or denizens of moist soil. *Marchantia* is a liverwort with a body superficially like that of foliose lichens, even possessing cup-like structures on the surface, but the cups are not dark colored. They are filled with reproductive cells called gemmae that are dispersed by the wind. There are also male and female liverworts that produce umbrella-

Figures 6-3A, 3B, & 3C: Lichens demonstrate three major growth forms. Crustose lichens (**A**) grow in thin sheets, usually on the surface of rocks. Many are brightly colored with red and yellow pigments that shield the algae from the full force of the sun. Foliose lichens (**B**) are more loosely attached to rocks, soil, or wood and look somewhat "leafy" in appearance. Fruticose lichens are upright on stalks, thread-like or straplike. Cladonia (**C**) often carpets forest floors and may be found growing among mosses.

like structures that you will not find on any lichen. Microscopically they are very different. Liverworts are made up of rather typical green plant cells, whereas lichens are basically a fungus with a layer of algal cells trapped near the surface.

Fungus Incognito

Basically, a lichen is a fungus in disguise. Lichen fungi are transformed by the algae they "embrace," somewhat like a plant stem that creates a gall in response to an insect invasion. This transformation results in a unique lichen body, yet the basic growth form you see and the reproductive structures that adorn it are unmistakably fungal in origin. Most of the fungi that form lichen associations are Ascomycetes, also called sac fungi because their spores are carried in sacs called asci. Ascomycetes, like the Basidiomycetes we looked at in the last chapter, are a major division of the fungus kingdom.

Figure 6-4: Moss and lichen often are found together. Lichens are stiff and usually gray or mineral green in color. Mosses are soft with bright green whorls of tiny leaf-like structures. Sporangia are borne on long stalks.

Find a foliose lichen with black- or brown-lined, cup-shaped structures on its surface. You have discovered a lichen with an Ascomycete fungal partner. The "cups" are apothecia and the dark linings contain fungal spores. However, if you were to cut through the plant body and look at it in cross section, you would see the uniquely lichen organization. A layer of green underlays a clear outer cortex. The green layer is made up of individual algal cells held in a meshwork of hyphae. Below that you may see loosely entwined hyphae with no algae. This layer is called the medulla. Completing the "sandwich" is a layer of somewhat more fibrous cortex from which root-like rhizinae project. In one major grouping of lichens the medulla doesn't exist, and the algal cells are dispersed among the hyphae between the two cortical layers.

Marchantia

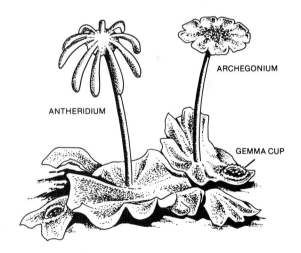

Figure 6-5: The liverwort Marchantia *is found in low-lying moist areas. Male and female gametes are borne on distinctive stalks. Gemma cups contain gemmae—special capsules that are washed away during rains and can generate new plants asexually when conditions are favorable.*

The Lichen Plant Body

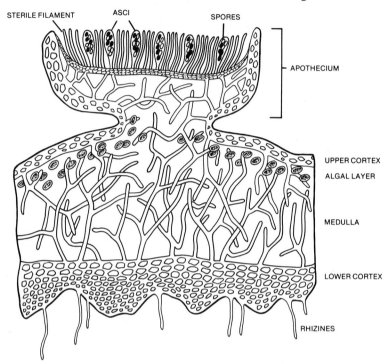

Figure 6-6: *A cross section through a lichen plant body and apothecium would resemble this illustration. The upper cortex is clear, allowing light to reach the algal layer below. The medulla is a layer of twisted, intertwined fungal hyphae. Fungal spores released from the apothecium must encounter free-living algal cells to form a new lichen.*

Other fungal reproductive structures that you will find in lichens are perithecia and pycnidia. Perithecia are flask-shaped, spore-producing structures that are sunk into the lichen tissue. They appear as dark spots on the lichen's surface. Pycnidia are similar to perithecia, but smaller, and produce tiny cells similar to the male sex cells of some fungi. Although there is no evidence for sexual reproduction be-

tween lichen fungi, pycnidia are common, and it wouldn't be too surprising to hear of lichen sexuality some day.

The spores produced by apothecia and perithecia fall to the ground, where they must encounter the right kind of algal cells (usually a species called *Trebouxia*) in order to form a new lichen. Lichens do have a few ways to reproduce, however, that are less "chancey." If a lichen is physically broken up, each piece can generate an entire new plant. This process of fragmentation is common in simple plants (and even animals). Knob-like growths, called isidia, are a more unique method of lichen reproduction. Each isidium consists of a bundle of hyphae with its own complement of algal cells. The isidium is covered by a layer of cortex. When the isidia are brushed off by a passing animal or the forces of wind and rain, they fall to the ground possessing all the parts needed to create a new lichen plant. Similarly, some lichens have smaller structures that collectively look like a fine powder on the thallus. These soredia are ball-shaped collections of hyphae and algal cells that also can start entire new plants.

The algal partners in the lichen association usually reproduce by asexual cell division. Rarely, one might witness sexual reproduction (see Ahmadjian, 1982). In any case, there is no provision for dispersing algae independently of the fungus. However, it is possible to separate the component parts of a lichen in the lab. You may want to give it a try.

Taking Lichens Apart . . .

To take lichens apart we can exploit their reproductive techniques. Find a foliose lichen with large apothecia. If the lichen is dry, wet it. Do you notice any color change? As the algae absorb water, they often turn a darker green. Lichens survive dry and hot spells by shutting down activities—taking extended physiological siestas. When wet, they undergo a surge in both photosynthesis and respiration. Take a razor blade and gently scrape the surface of the lichen. When you scrape away the top thin skin of fungus, a bright green layer of algae should be visible beneath. Take some of this green material and place it on a glass slide. If you have a microscope you can place a drop of water on the preparation and look at it under

Figure 6-7: Perithecia are similar in structure to apothecia but are sunk below the surface of the plant body. Externally they appear as black spots where they open to the surface.

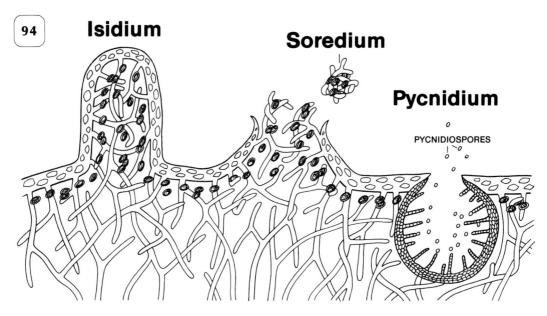

Isidium

Soredium

Pycnidium

PYCNIDIOSPORES

Figure 6-8: *Isidia and soredia are uniquely lichen reproductive bodies that disperse fragments of fungus and algae together. Isidia are knob-like growths that are broken off by wind, water and animals. Soredia are carried away like a fine powder by the wind. Pycnidia are simple spore-producing structures common among fungi. The pores that open to the surface are smaller than those of perithecia.*

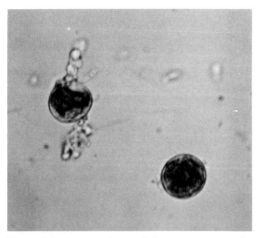

Figure 6-9: *If you moisten a lichen, scrape away the cortex and place a sample of the algal layer on a slide, you will see individual algal cells, many of which have fragments of hyphae still attached to them.*

medium to high magnification. You will find many small, usually round algal cells, often with fragments of fungal hyphae still attached.

If you don't have a microscope, take some garden soil, put it in a small dish, and wet it. Put some of the algal cells on the soil and cover them with a small piece of glass or transparent sandwich wrap. Keep the dish under a growlux-type light (they have the same spectrum as sunlight). Under constant but low-intensity lighting conditions, algae will tend to outgrow fungus and form a greenish layer on the soil. Alternatively, you could try a similar experiment under more controlled conditions by making a mineral solution or a nutrient agar in which the algae can grow (see Appendix III).

The fungal partner can be isolated in the following way. Prepare a culture medium of 2 percent agar or soil-water medium (see Appendix III). Wash a sample of the lichen plant body under tap water. Take a small piece that has an apothecium on it and attach it, apothecium down, to the top of a culture dish with some petroleum jelly as illustrated in Figure 10. As the apothecium drys out, spores will fall to the agar surface. Typically, 50 percent will germinate. Germination time may be from a few minutes to a few weeks. Crustose forms seem to germinate faster than foliose.

... And Putting Them Together Again

As a general rule I've found it's harder to put things together than take them apart. This rule applies to lichens, also. In fact, it's only been in the last few years that scientists have had consistent success at getting algae and fungi together to form lichens. Part of the problem is that lichens are very slow growing. It may take four to six months in laboratory culture to get resynthesized lichens to produce reproductive structures which, even then, may

be sterile or immature. There was also some confusion about the conditions required for lichen synthesis. Some thought alternate wetting and drying was necessary, but that seems not to be the case. In fact, a constant 90 to 95 percent humidity seems to work best. You also have to get the right alga and fungus together or the latter will eat the former. Some of the natural bacteria and viruses in a lichen's environment may also contribute to the development of healthy lichens.

If you'd like a challenge and want to try to resynthesize the lichen you took apart, here's a technique that worked for Ahmadjian and Jacobs (1981): Take 125-ml flasks, similar to the one illustrated in Figure 11. Put 20 ml of a 2 percent agar solution into each. Stopper the flasks with non-absorbent cotton plugs. Place a cleaved mica strip, 1 x 7.5 cm, that has been soaked in a solution of Bold's Basal

Germinating Ascospores

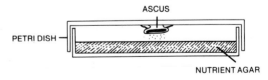

Figure 6-10: You can separate the fungal partner of a lichen by securing an ascus to the underside of a petri dish cover with petroleum jelly and allowing it to dry out over a layer of nutrient agar. Hyphae will grow from the germinating ascospores.

Medium (see Appendix III) into each flask so that one end is embedded in the agar and the other rests against the side of the flask. Take samples of fungus that have been growing in liquid medium, wash them, and put the samples into petri dishes where they can be dried with filter pads. Add algae from agar cultures to the fungus and press the mixture onto the mica strips. Stopper the flasks with cotton plugs and cover the cotton with aluminum foil. Give the flasks a regular day-night regime and give the lichens plenty of time to grow.

If we could accelerate the events that occur when alga and fungus meet, they would seem much more dramatic. The lichen association might best be described as a controlled parasitism, with the fungus being the aggressive partner. It is only when an alga can partially resist the invading fungus with chemical defenses that a partnership can form. As the mat of fungus dries out on each mica strip, it acquires a doughy consistency. From its pock-marked surface, strands of fungal hyphae rise slowly into the air. As the algal cells dry, they secrete a sticky, gelatinous material. While this secretion is probably a technique for conserving moisture, the gooey matrix succeeds also in sticking to the exploring hyphae.

The strands of fungal tissue coil like pale snakes around whatever small objects

A Method of Growing Lichens From Their Individual Partners

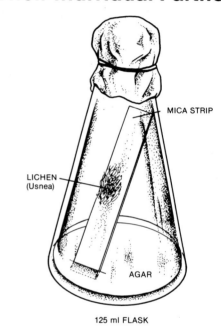

MICA STRIP

LICHEN
(Usnea)

AGAR

125 ml FLASK

Figure 6-11: *Lichens have been successfully reconstituted (although it is difficult) by allowing the separated algae and fungus to reassociate on strips of mica. See text for details.*

they meet. When they encounter algal cells, however, they may do several things. If the algal cells have no defense, the fungus will penetrate and consume them like any other food item on the fungal menu. But if a lichen association is to form, the fungal hyphae flatten into padlike structures called appressoria where they contact the algal cells. The hyphae resume their normal shape beyond the point of flattening and may "hop" to another cell or another point on the same cell. Alternatively, or in addition to appressoria, the fungus produces finger-like extensions called haustoria that indent the algal cell like a finger poked into a balloon. Both appressoria and haustoria most likely serve to increase the surface area shared by fungus and alga and thus increase the efficiency of food flow between the two.

Bundles of algae form, surrounded by intertwining strands of fungus. These structures resemble the granular soredia we looked at earlier. As lichenization progresses, another network of hyphal strands slowly ties the soredia together. At this point the fungus begins to secrete a gelatinous matrix of its own which cements the two organisms into a common "body." Fungal hyphae begin to differentiate, becoming dry and leathery on the outside and forming a more loosely woven medulla beneath a now distinct layer of algal cells. In nature, viruses, bacteria, mites, nematodes, and probably many other creatures will become involved in some way with the association.

When (and if) reproductive structures form in your cultures, the transformation is complete. Where two organisms were struggling to survive in a dry, nutrient-poor environment, there is now one composite creature much more likely to survive.

A Chemistry for Success

Lichens are survivors.

You can find them farther north and farther south than most plants. You can find them high on mountains, in dry deserts, and even within rocks. The lichen *Verrucaria serpuloides* is continuously submerged in the cold coastal waters of Antarctica. *Lecanora esculenta*, a so-called "wandering lichen," is blown by the dry winds of some deserts. As we've already seen in the first chapter, lichens discovered in 1974 live quite comfortably inside Antarctic rocks that suffer temperature ranges from +30°F to -158°F in winter.

At least part of the lichen's success stems from its unique biochemistry. One aspect of this biochemistry is lichen physiology. Put quite simply, lichens slow down when the going gets tough. Photosynthesis, the process whereby green plants combine carbon dioxide and water with the energy of sunlight to produce sugars and oxygen, is a two-step process. The first step produces a modest amount of energy. Some photosynthetic bacteria get along nicely without ever going to step two. So do lichens, under adverse conditions, but this low-energy strategy allows for little, if any, growth. When water is plentiful and temperatures are moderate, however, the algae in lichens, like most green plants, kick in the photosynthetic afterburner and complete step two. Using the products of the first step, along with the energy-gathering abilities of several kinds of chlorophylls and other pigments, the algae produce more high-energy compounds (ATP, for example) and oxygen. However, this second step in photosynthesis bleaches the chlorophylls and uses up more raw materials, so there is a price the lichen pays when it lives in the energetically "fast lane."

In good times, lichens draw on food reserves that are primarily carbohydrates: sugars and starches. Lichen algae commonly produce the sugar alcohol, ribitol, much of which is exported to the fungus. Under stress, lichens store fat in the form of droplets called pyrenoglobuli. Similar tactics were discovered in the phytoplankton of Antarctic waters, where 80 percent of the carbon fixed during photosynthesis is in the form of fat droplets.

When lichens are moistened after a period of dryness, they tend to crank into gear quite quickly. Like jackrabbit starts at the stop sign, this "quick start" tends to burn energy. Carbohydrates are produced in excess of need, and the respiration rate is higher than normal. However, lichens that are adapted to different temperature and moisture conditions seem to be able to regulate their maximum-efficiency thermostat. This is measured in terms of the amount of carbon dioxide the algal partner can assimilate. Scientists are un-

sure whether the same fungus meets up with different strains of algae in different environments or whether one kind of alga can alter its physiological rate.

Lichens also produce a unique series of chemicals called lichen substances, which may play several important roles in their biology. One such chemical, usnic acid, is implicated in the control of the flow of nutrients from alga to fungus. Algal members of a lichen partnership, when isolated from the fungus, almost immediately stop exporting ribitol. When part of the lichen association, on the other hand, they give up some 40 to 90 percent of their carbohydrates to the fungus. One theory to explain this involves usnic acid and urea in an interesting feedback loop. Urease, which is an enzyme commonly found in lichens and their fungal partners, promotes the breakdown of urea into carbon dioxide and ammonia. Ammonia tends to increase the respiration rate in the algae, and the carbon dioxide stimulates photosynthesis. Increased photosynthesis results in usnic acid formation. Usnic acid, in turn, inhibits the action of urease. Urease and usnic acid thus regulate each other and, in the process, regulate rates of respiration and photosynthesis.

Usnic acid and other lichen substances are also antibiotics. They inhibit the growth of bacteria. You might speculate that this could help them ward off infections by various bacteria in their environment. Surprisingly, though, the strains of bacteria they inhibit—those important to man—are not usually the kind commonly found in soil.

Because lichens are often pioneer plants on bare soil and rock, some speculate that lichen acids are chelating agents: chemicals capable of binding metals and freeing them from combination with other minerals. This chelating ability would make the metals available to the lichen.

Although some lichen acids are colorless, others come in shades of bright yellow and orange. These colors are particularly conspicuous in crustose lichens that are almost always exposed to direct sunlight. The heavy pigmentation seems responsible for shielding the delicate algae from overexposure and thus the bleaching of their chlorophyll.

Chemical Fingerprints

Whatever their function in lichen survival, lichen substances provide a service to the naturalist. Because many of them are unique to lichens or to certain fungi which form lichens, they serve as chemical I.D. tags for various species. Certain simple tests can be applied to lichens with three reagents: calcium hypochlorite (bleaching powder), which you can abbreviate as C; potassium hydroxide (caustic lye), abbreviated K or KOH; and paraphenylenediamine (P). Calcium hypochlorite is a white, pungent powder that must be mixed fresh with water each day. Many commercial bleaches can be substituted for the pure reagent, as calcium hypochlorite is their active ingredient. Potassium hydroxide is available from supply houses in pellet form. The containers should be tightly closed to prevent the pellets from picking up water from the air. Handle this caustic reagent with gloves or tweezers. Dissolve in water to make a fairly concentrated solution. In a well-stoppered bottle it should be stable for months. Paraphenylenediamine is a dark powder. Dissolve a small quantity in 5-10 ml of ethyl alcohol (rubbing alcohol or even acetone will also work as the solvent). This reagent is also available from supply companies, but may be harder to get.

Use these reagents to perform color tests on lichens. To test the lichen, carefully scrape away part of the upper cortex with a razor blade to expose an area of white medulla about two millimeters square. You may want to use a hand lens. Use a fine pipette or medicine dropper to apply a reagent. Note any color changes. The color reaction is usually unmistakable, but learn to distinguish a spurious yellow color sometimes caused when

Crystal Test

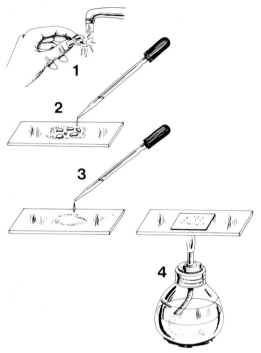

Figure 6-12: *Organic acids and other lichen substances can be dissolved from lichen plant bodies (thalli) by an organic solvent like acetone. The substances are recrystallized by other reagents and the form of the crystals are used to identify the substances.*

1) Take a piece of lichen thallus and clean it under running water. Break up the thallus fragment into very small pieces with tweezers and place a few fragments on a glass slide.

2) Add a drop of acetone. When the first drop dries, add another. If a whitish residue is left when the acetone dries, a lichen substance is present. You may want to add another drop or two of acetone to dissolve out more of the substance. When the acetone is dry, carefully brush off and discard the lichen fragments.

3) Add a drop of recrystallizing agent to the slide. (See text for a list).

4) Add a coverslip and heat over an alcohol burner until bubbles begin to form. When cool, examine under the microscope and compare the crystals with those in Figure 6-13 or photographs in How to Know the Lichens by Mason Hale.

REAGENTS

G.E. (glycerin-acetic acid, 3:1)

G.A.W. (glycerin-95% alcohol-water, 1:1:1)

G.A.o-T. (glycerin-alcohol-o-toluidine, 2:2:1)

G.A.An. (glycerine-alcohol-aniline, 2:2:1)

G.A.Q. (glycerine-alcohol-quinoline, 2:2:1)

KOH is added to the upper cortex. This is caused by the algae absorbing moisture.

The color changes induced by the several reagents imply the presence of certain lichen substances. The presence or absence of these substances, in turn, is used in keys to distinguish one species from another. The best lichen key is the book *How to Know the Lichens* by Mason Hale. This book will provide more detailed information on chemical tests, but the chart on page 101 gives you an idea of what you can tell from the tests.

You'll notice that the color tests don't narrow down the lichen substances completely. For a more positive identification you need to perform another chemical test and examine your results under a microscope. The effort is worth it, because the lichen substances are converted into elegant crystals. Some are delicate rosettes, others are long, curved needles arranged in haystack shapes, and still others are tangled sprays that remind you of nerve networks. Under polarized light a

Pigmented substances

REAGENT REACTION	SUBSTANCE POSSIBLY PRESENT
Orange or red pigments K+ (purple)	parietin, rhodophyscin, solorinic acid
Yellow pigments K- (or yellowish)	calycin, pinastric acid, pulvic acid, usnic acid, vulpinic acid

Colorless substances

K+ (yellow, or yellow turning red)	atranorin, baeomycic acid, galbinic acid, norstictic acid, physodalic acid, salacinic acid, stictic acid, thamnolic acid.
K- (or brownish) P+ (yellow, orange or red)	fumarprotocetraric acid, pannarin, protocetraric acid, psoromic acid.
K-, P-, C+ (pink or red)	anziaic acid, gyrophoric acid, lecanoric acid, olivetoric acid, scrobiculin.
K-, P-, C+ (green)	didymic acid, strepsilin.
K-, P-, C-, KC+ (pink or red)	alectoronic acid, cryptochlorophaeic acid, glomelliferic acid, lobaric acid, norlobaridon, physodic acid
K-, P-, C-, KC-	barbatic acid, bellidiflorin, caperatic acid, diffractaic acid, divaricatic acid, evernic acid, grayanic acid, homosekikaic acid, lichexanthone, merochlorophaeic acid, perlatolic acid, protolichesterinic acid, rangiformic acid, sphaerophorin, squamatic acid, tenuiorin, ursolic acid, zeorin.

rainbow of colors emerge that would please the eye of an abstract painter.

For the crystal test you need acetone to dissolve the substances out of the lichen, a crystallizing reagent, microscope, microscope slide and coverslip, and an alcohol burner. Figure 12 outlines the steps for performing a test. You can try one or more of the reagents listed on page 100.

After the crystals form, compare them to photographs for identification. Again, *How to Know the Lichens* is a good place to start. A few of the different crystals are shown in Figure 13. The main reasons for failure in this test are: 1) too much reagent is used, so that the coverslip floats; 2) overheating dissolves all of the residue; and 3) the residue is too small to re-crystallize.

To see the crystals with polarized light you can buy small pieces of polarizing material inexpensively from places like Edmond Scientific. Cut two pieces of the material, one to fit over the substage light source and the other to hold over the eyepiece. As you look through the microscope, rotate the polarizing material over the eyepiece. At certain positions you will see beautiful and varied color effects as the polarizing material and crystals combine to allow only certain wavelengths of light to reach your eyes.

Lichen Relationships

In the far northern regions of the world lichens are at the base of the food chain. Extensive areas of ground are covered by dense mats of *Cladonia* and *Stereocaulon*, which serve as the primary food for caribou. In your neighborhood, lichens may not be as prevalent or as prominent in the food chain, but still play important roles in local ecology. In the mountains of northern Colorado, near where I live, I recently picked up what I thought was a fragment of lichen. It turned out to be a whole community. There was the gray, strap-like thallus of the lichen, of course, but it overlay a patch of moss. The moss held on to a wad of soil and pine needles, while its green, leaf-like parts poked between the lobes of the lichen. On the surface of the lichen a tiny red mite ran for safety. Under the dissecting scope the lichen fragment was a fairyland of hills and valleys, pitted, gray surfaces shadowed by moss fronds and decorated with the shed carapace of some small insect.

Naturalists, perhaps better than many other people, are especially attuned to the fact that no organism lives alone. A corollary to this truism might be that an organism's ties to other creatures in the organic network of our world are often complex, intricate, and unexpected. Lichens, for example, have an important role in the nitrogen economy of many habitats—especially in the tropics. This role develops out of many lichens' association with bacteria and blue-green algae that are capable of "fixing" atmospheric nitrogen into a form, ammonia, that green plants can use. Some blue-green algae are parasites on lichens, some are full-time symbionts, and others can form warty or knobby bodies, called cephalodia, that almost appear like grafts on the lichens. In the case of cephalodia,

as in the typical lichen relationship with green algae, the fungus gets the best of the deal. Virtually all the fixed nitrogen goes to the fungus. But when the lichen dies, nitrogen "fertilizers" are added to the local ecology.

Lichens also have a remarkable ability to absorb trace minerals and nutrients that contact their surfaces. In forests and other habitats where lichens often live epiphytically on other plants, this ability allows them to make use of rare minerals dissolved in rain water. In industrial societies the capacity to absorb chemicals can lead

Lichen Substances

USNIC ACID

ALECTORONIC ACID

CRYPTOCHLOROPHAEIC ACID

CAPERATIC ACID

OLIVETORIC ACID

GRAYANIC ACID

Figure 6-13: A few common lichen substances are illustrated. Refer to How to Know the Lichens by Mason Hale for other examples.

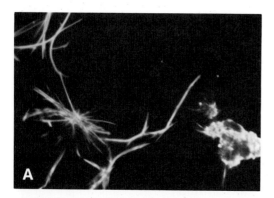

Figures 6-14A & 14B: *Lichen substances are striking under polarized light. Olivetoric acid is shown in* **A** *and Usnic acid in* **B**.

to their undoing because they accumulate sulfur dioxide, radioactive ions, heavy metals, and other potentially toxic materials. In the 1950's, scientists began to recognize that lichens could serve as good pollution monitors. Many species are very sensitive to sulfur dioxide, for example.

Other species, like *Lecanora conizaeoides* in England, have managed to adapt to high-pollution areas—and perhaps even use them to their advantage. Recently, Mason Hale, a Smithsonian lichenologist, has surveyed lichens in Wyoming for trace metals and found that lead levels are three times higher than expected—an indication that acid rain and other "fallout" from industrial society is a worldwide, not just a regional problem.

Fortunately for lichens, they don't produce anything that humans find necessary or irresistible. They were used for dyes for much of human history, but modern man has synthetic substitutes that are cheaper, longer lasting, and easier to manufacture. Lichens are so slow growing that their lichen acids are not worth harvesting as antibiotics. I once read a science fiction story, *The Trouble With Lichen,* where a scientist found the key to immortality in a lichen. If that should come to pass, lichens might be in trouble for a while, but only if humans succeeded in decimating lichens before knocking off each other for the secret.

Much about lichen ecology remains to be learned. There is the question, for example, of why some lichens produce substances that kill bacteria not normally found in their environment. Toadstools and other fungi produce toxins that kill nematodes and other small animals. Do lichens have such an arsenal? A sugar called trehalose has been found important for certain creatures that survive long periods of dessication. The molecules of the sugar somehow fill in intercellular spaces and prevent delicate membranes from damage. Does trehalose play such a role in lichens? The answer to these and other questions may eventually lead to an understanding of how plant life initially bridged the gap between the world of water and the world of dry land. Just as Neil Armstrong couldn't have taken his "giant leap" without a whole lot of energy being expended by others, the first "invaders" of dry land constituted a consortium of creatures that, together, could utilize the new habitat. Also, like our first landing on the moon, the first invasion or two probably didn't "take." But the example of life on Earth has been that nearly any kind of creature can arise. It's only a matter of time . . .

REFERENCES

Ahmadjian, V. 1967. *The Lichen Symbiosis.* Waltham, Massachusetts: Blaisdell.

Ahmadjian, V. and J.B. Jacobs. 1983. Algal-Fungal Relationships in Lichens: Recognition, Synthesis, and Developments. In *Algal Symbiosis,* Lynda J. Goff, ed. New York: Cambridge University Press.

Ahmadjian, V. 1982. Algal/Fungal Symbioses. In *Progress in Phycological Research,* Vol. 1, Round/Chapman, eds. Amsterdam, Holland: Elsevier Biomedical Press B.V.

Crosscurrents Section: Litmus Lichens. 1986. *Science 86,* vol. 7, no. 1 (January-February).

Deason, T.R. and H.C. Bold. 1960. *Phycological Studies I, Exploratory Studies of Texas Soil Algae.* University of Texas publication #6022.

Difco Laboratories. 1972. *Difco Manual of Dehydrated Culture Media and Reagents,* 9th ed. Detroit, Michigan: Difco Laboratories, Inc.

Hale, Mason E. 1967. *The Biology of Lichens.* London: Edward Arnold.

Hale, Mason E. 1969. *How to Know the Lichens.* Dubuque, Iowa: Wm. C. Brown.

Lilly, V.G. and H.L. Barnett. 1951. *Physiology of the Fungi.* New York: McGraw Hill Book Co.

Raham, R. Gary. 1978. Exploiting the Lichen Liaison. *The American Biology Teacher,* vol. 40, no. 8 (November).

Science 86. 1986. Crosscurrents Section: Litmus Lichens. *Science 86,* vol. 7, no. 1 (January-February).

Scott, D. 1969. *Plant Symbiosis.* New York: St. Martin's Press.

Smith, D.C. 1973. *The Lichen Symbiosis.* New York: Oxford University Press. Available in Oxford Biology Reader series from Carolina Biological Supply Co., Gladstone, Oregon.

Smith, D.C. 1973. *Symbiosis of Algae With Invertebrates.* New York: Oxford University Press. Also an Oxford Biology Reader.

7

The Cambrian Explosion

It was a time of monsters. *Hallucigenia* ambled about on seven pairs of stilt-like spines looking for carrion. Seven tentacles sprouted from its slender back and twisted with the currents. The five-eyed *Opabinia* moved sinuously along the bottom, raising clouds of silt while its single forward-pointing tentacle, tipped with a ragged claw, searched for worms. However, it was also a time for vaguely familiar forms, somehow distorted, as if seen through an imperfect lens. Many of the worms were not unlike marine forms you can see today. Sponges were plentiful and there was a host of arthropods, recognizable by their hard skins and jointed legs.

The time was the geological period called the Cambrian, and the place was 530 feet below the top of a shallowly submerged reef. Over 119 genera of animals made their home above, on, or under the muddy bottom. These creatures were representatives of a diversification of animal life that later would be called an explosion because of the apparent speed with which nearly all the major groups of animals suddenly appeared in the fossil record. Some of this rapid development is an illusion caused by the rarity of fossils of soft-bodied animals. The development of shells and other external skeletons, however, may have been an innovation that initiated many experiments in body plans.

The animals and plants involved, of

Figure 7-1: *This reconstruction of the seafloor community that would later become the Burgess Shale fossils shows some of the species from several phyla of animals. The five-eyed, tentacle-snouted form near the center called* Opabinia *has no known living relatives. It is shown grabbing* Hyolithes, *one of the few molluscs found at the site. A priapulid worm rears out of its burrow. Its living relatives look much the same today. The stalked Dinomischus, like* Opabinia, *belonged to a phylum now extinct. In the foreground, a many-legged arthropod with large "horns" scuttles over a rock while the scaled and spiked* Wiwaxia, *a possible mollusc, plows through the sediments. The tree-like forms are sponges of the genus* Vauxia. *(Drawing based on illustrations and information from Morris, 1979 and Morris and Whittington, 1979)*

course, were just making a living as best they could. Occasionally their existence was interrupted by periodic disasters. The

The Burgess Shale Geography

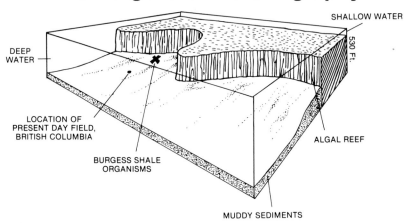

Figure 7-2: The Burgess Shale organisms lived in the muddy sediments at the base of an algal reef that rose 530 feet above them. The sea floor "slumped" periodically, sliding the inhabitants into cold, deep water that helped preserve their remains. (Redrawn from Morris and Whittington, 1979)

sloping bottom below the reef would give way and sweep whole communities swiftly away, casting them into deeper waters, cold and without oxygen. There the bodies decayed slowly, if at all, and successive layers of mud crushed them flat, ultimately compressing a foot of sediment into a life-stained inch of shale.

Over five hundred million years later a man on horseback would stop to rest his horse that had lost its footing on a trail near Field, British Columbia. Being a geologist, he picked up an interesting piece of gray rock and cracked it open. Within lay a beautifully preserved invertebrate—the first specimen from the Burgess shale. The man was Charles Doolittle Walcott, Secretary of the Smithsonian Institution. It was the fall of 1909, too late in the year to seriously look for the source of the shale, but the next season he returned with his sons and they located the outcropping they sought high in the Rockies near a glittering green lake.

Several expeditions to the Burgess shale, scattered over the decades since then, have revealed a unique look at the origins of some of the soft-bodied life you can now unearth in your gardens. Earthworms, slugs, snails, and roundworms are just some of the fallout from this Cambrian explosion, while creatures like *Halluci-genia* and *Opabinia* left no descendents for us to contemplate.

A Preference for Worms

Animal life on Earth has mostly been an experiment with worms. Approximately half of the major divisions (phyla) of animals could best be described as worms. Our perspective is distorted, of course, because two phyla tend to dominate our lives: the Chordates, or animals

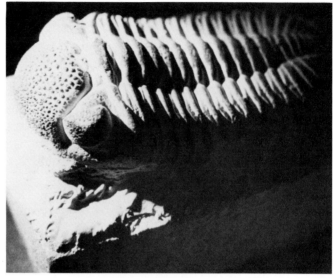

Figure 7-3: *Trilobites were a very successful group of arthropods that arose during the Cambrian and survived into the Triassic. Their segmented body plan is evident in this picture of a cast of* Phacops rana, *a trilobite from the Devonian silica formation of Sylvania, Ohio. Horseshoe crabs are probably their closest living relatives.*

with backbones, such as ourselves and all the other large animals that share our world; and the Arthropods. This latter group, particularly the insects, are so successful in terms of sheer numbers that it would be difficult to ignore them.

Nevertheless, the worm body plan is a basic one, perhaps because it is the simplest expression of two very animal characteristics: bilateral or mirror-image symmetry and cephalization. Of the 32 animal phyla, 28 display mirror-image

symmetry, which means that you can divide them into equal halves in only one way, resulting in halves that are equal but opposite in orientation—like the reflection in a mirror. Cephalization simply means that animals have a head and a tail end. Virtually all animals with bilateral symmetry are also cephalized.

Animals that share these common denominators are further divided into three groups based on characteristics that are not quite so obvious. Animals all start out

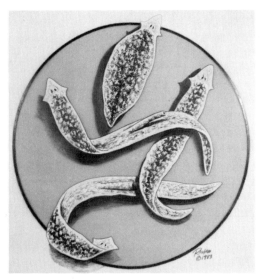

Figure 7-4: *The freshwater flatworm,* Planaria, *demonstrates two ancient and basic characteristics of complex animals: bilateral symmetry and cephalization.* Planaria *can be divided into two mirror-image halves by drawing a line down the length of its body from the center of the head to the center of the tail.* Planaria *has a recognizable head end containing a simple brain. Its "cross eyed" appearance is easy to recognize under low magnification.*

as a ball of developing cells called a blastula. In the more complex animals, the blastula soon specializes into three kinds of tissues which, from the inside out, are called endoderm, mesoderm, and ectoderm. Endoderm develops into the digestive tract and its associated organs, mesoderm becomes muscles and skeleton, and ectoderm ends up as nervous tissue and skin. The details of develop-

ment for the three groups differ, however, in the matter of body cavities.

Body cavities are called coeloms, a term scientists fashioned from the Greek word *koilos*, which means hollow. Some animals, like flatworms, have no body cavity, and they are described as being acoelomate. Flatworms and their relatives represent a rather small fraction of living animals.

Animals with body cavities, or coelomates, all have hollows in their body which result when the mesoderm splits and separates during development. A third group, the pseudocoelomates, have body cavities, but they are not surrounded by mesoderm, as proper coeloms should be. Rotifers, gastrotrichs, roundworms, and other worm phyla are pseudocoelomates. This group has been quite successful. There are 80,000 described species of roundworms, for example, (not to be confused with earthworms or other annelids) and probably 900,000 undescribed ones. A billion individuals per acre have been counted in the top two centimeters of rich garden soil. If all life except roundworms disappeared, you might well see a spectral image of the vanished forms created by their roundworm parasites.

Nevertheless, most of the animals you see in your gardens belong to the coelomate group. This group divides into two large divisions that were distinct at the time the Burgess shale organisms were swept away in their landslide. I mentioned that animals start out as a ball of cells called the blastula. It isn't long before this ball indents at one end to form a gastrula. The indentation in one group, the deuterostomes, ultimately becomes the anus. The indentation in the proterostomes becomes the mouth. Scientists have decided this is rather a fundamental difference. You are a deuterostome, along with other Chordates, starfish, and a couple of other small phyla. Arthropods, mollusks, and annelids, as well as many worm-like phyla, belong to the very successful proterostome group.

The Segmented Lifestyle

Two very successful Proterostome groups explored contrasting lifestyles. The mollusks, which I'll talk about later, developed (and later lost again, in some cases) the shell and acquired distinctive methods of locomotion and feeding. The annelids, however, explored the possibilities inherent in segmentation. In essence they are composed of nearly identical subunits which duplicate many body systems. We and other chordates retain remnants of this segmentation which is particularly evident during early stages of development, but is also illustrated by such things as the sequence of vertebrae that form our backbone.

The earthworm is an annelid worth studying, even though it is atypical. The most common annelids are marine bristle worms called polychaetes. The genus *Canadia*, found in the Burgess Shale, has bunches of feathery bristles used in locomotion and nearly identical to those of present-day bristle worms. Leeches are also annelids, along with a small group of worms that parasitize starfish and their relatives. Earthworms are readily found, however, and they have an important role in the ecology of land-based communities. Charles Darwin went so far as to say, "It may be doubted if there are any other animals which have played such an important part in the history of the world as these lowly organized creatures."

Although earthworms are usually drab in color, dark brown on top and lighter brown underneath, many marine annelids are brightly colored, often striped or spotted montages of pink, brown, or purple. Earthworms take on a pinkish tint because their blood contains hemoglobin, as ours does. Some British species are greenish because the oxygen carrying pigment is the green chlorocruorin. You can see the main blood vessel as a colored line just beneath the skin on the top (dorsal) side of the animal.

Annelids breathe through their body wall, via gills, or through paired appendages called parapods that may also be used for locomotion. They have a brain of sorts—basically a collection of ganglia at the front end—and a nerve chord which runs along the belly (ventral side).

Annelids are sensitive to light, relying on eyes or light-sensitive cells scattered over their skin. They also have organs of balance called statocysts. They show their relationship to mollusks by sharing early developmental stages that are nearly identical; both have trochophore larvae (see Figure 7-9).

Earthworm Biology

Why did Charles Darwin have such a high opinion of the earthworm's importance? Charles Frederick Holder in an early biography of Darwin claims that Darwin's uncle encouraged him to think about the abilities of earthworms. He suggested to Darwin that the sinking of stones on his farmland might be due to the constant tunneling of earthworms. In 1842 Darwin began an experiment that wouldn't be completed for 29 years. He covered one part of a field with broken chalk and another part with cinders. In November of 1871 he dug a trench across the field and found both the chalk and cinders to be in a layer seven inches below the surface. He concluded that the soil had been turned over at a rate of .22 inches per year.

You can find the mounds of black earth, or castings, that have been through the earthworm's digestive system on the ground or on lawns. Since there may be some 50,000 worms per acre of ground, they can transfer 18 tons of castings to the surface in a single year. Earthworm activities grind up the soil, reducing it to a finer texture; fertilize the soil by secreting lime which neutralizes its acidity; and enrich the soil by burying bones, leaves, and other organic matter where it can decompose and provide soil with drainage and aeration.

Collect a few earthworms and let's do a firsthand inspection. Put them in a shallow pan with a few moist paper towels. You can tell which is the front end by the direction they move. Notice that the front end is more pointed. The mouth is covered by a lip called the prostomium. If you want to see how the prostomium works to aid feeding, offer the animal a scrap of lettuce or cabbage leaf. Try tilting the pan. How does the worm respond to the pull of gravity? Put the worm on the dry side of a towel that is moist on one end. Which way does it move? Worms are dependent on a moist habitat because they breathe through their skins rather

Earthworm Anatomy

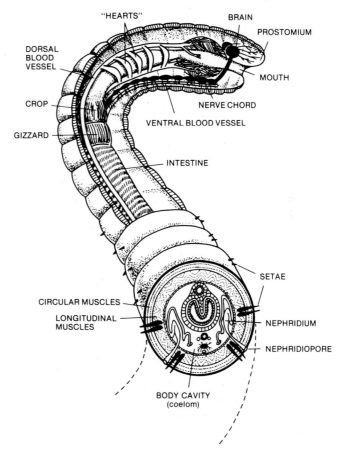

than with lungs.

Were your earthworms hard to pull out of the ground? If you look at the segments with a hand lens, you will see four pairs of stiff bristles (setae) on every segment except the first three and the last. They use these bristles to anchor themselves in their burrows and can hang on quite tenaciously. Set the earthworms on bare ground and you can see how they use the bristles to move forward. They extend the anterior part of their body and extend the bristles there to hook the ground. Rear setae are relaxed, and the posterior end is drawn forward. The latter setae are then extended and the anterior ones retracted so that the head end can be moved forward again. Strong longitudinal and circular muscles coordinate the movement, acting against the pressure of coelomic fluids much as if they were a skeleton.

If you look part way between the lateral and ventral (belly-side) bristles with a hand lens, you can see a small pore in all

Figure 7-5: One of the most obvious characteristics of annelids is their segmentation. Many organs and structures are replicated in each segment. Anterior segments show some specialization because the nervous system and mouth are located there. Annelids also have their nerve chord along the belly (ventral) side and their major blood vessel on the back (dorsal) side. They share this characteristic with arthropods.

except the first three and last segments. This pore is the opening of a nephridium, an organ that serves as a kidney for the earthworm. There is a pair of nephridia in all except the four segments mentioned.

A pale, smooth ring is a conspicuous feature of earthworms and can usually be found between segments 32 and 37 (start counting from the prostomium). This structure, called the clitellum, will eventually form a cocoon for the development of eggs. Earthworms are hermaphrodites, a term derived from the Greek legend of Hermaphroditus, who, while bathing, somehow became bodily attached to the Nymph, Salmacis. Biologically speaking, however, hermaphrodites are organisms that produce both eggs and sperm. In the case of the earthworm, one individual can't fertilize itself, but must get together with another of its kind.

Earthworms mate in May. If you go out to your garden on a moist night with a flashlight, you can probably observe the nuptials. Mating pairs bind themselves together with a thick mucous and exchange sperm. The sperm are produced in swellings on the ventral side of segment 15. They are deposited in receptacles that open to the outside between segments 9 and 10 and 10 and 11. After the worms separate, their respective clitella secrete cocoons which begin to move forward. The cocoon slides over segment 14 and picks up a load of eggs. When it moves over segments 9 through 11, it picks up the sperm recently deposited there. The cocoon continues to move forward until it slides off the head to form a football-shaped object the size of an apple seed. The eggs become fertilized in the cocoon and incubate two to four weeks before young worms emerge. They become sexually mature in 60 to 90 days and are full-grown in about a year.

Fold over a corner of the paper toweling in your tray. Do the worms seem to prefer exposed or sheltered areas? Earthworms are sensitive to light, although they have no eyes as such. Photoreceptive cells in the epidermis are sensitive to various wavelengths of light. Some receptors respond to blue light, which send worms back to their burrows at dawn. Other receptors respond to yellow light. Earthworms don't respond to red light, however, so if you cover your flashlight with red transparent acetate you can catch them unawares.

Gently prod your earthworm with a toothpick. What parts of the animal are most sensitive to touch? Earthworms respond to vibrations and other stimuli via sensory cells connected to fine hairs that project through the cuticle. A thin, transparent membrane also covers the body and helps protect it from physical and chemical injury.

Teaching Earthworms New Tricks

Although earthworms have never ranked high in I.Q. tests, they are capable of learning to avoid noxious materials. Try an earthworm out in a "T" maze similar to those used in rat learning experiments.

Build the maze from wood, cardboard, or other convenient material. At one end of the T-bar place a sample of rich organic soil. At the other end place filter paper soaked in a 0.5 percent solution of ammonium hydroxide. Mark individual earthworms with India ink so that you can recognize them. Allow each worm to run the maze and record which way they turn when they enter the T-bar. Run up to 50 trials with each worm, making sure they don't dry out in the process. How long does it take for a given worm to consistently turn toward the humus? Are there individual differences? Do worms remember what they've learned after a lapse of a day or more?

Try variations on the noxious stimulus, substituting a dilute acid or perhaps covering that end of the T-bar with coarse

Figure 7-6: Snakes like this garter snake (Thamnophis sirtalis) find earthworms to be a tasty snack. If you put worm and snake into a terrarium you can watch the snake close in on the worm (even if the worm has "dug in"), strike, and gulp it down.

sandpaper. You could also test for sensitivity to various colors of light or place another earthworm behind a wire mesh partition to see if they prefer soil to another earthworm's company. Does an earthworm respond to another worm behind a glass, rather than wire, partition?

You're sure to think of other possibilities as you work with the apparatus. When you finish, you'll have a much better idea of how annelids might cope with their environments.

The Molluscan Strategy

Mollusks were fairly rare in the Burgess Shale community. Three genera have been identified, which represents just 2.5 percent of the animal life found there. *Hyolithes* had a cone-shaped shell with a protective cap at one end. Although at least one species of burrowing worm found the shell indigestible, these worms successfully preyed on this mollusk. *Scenella* was another bottom-dwelling mollusk and it possessed a shell that looked like a "Chinese hat." *Wiwaxia* was a creature whose molluscan heritage is a little more problematical, but it appears to have had a feeding organ similar to those of modern mollusks, and its body form, although adorned with long spines, was

rather similar to that of some primitive mollusks. If mollusks were rare 530 million years ago, however, they were merely at the beginning of a long and successful career. Today mollusks are second only to arthropods, with an estimated 100,000 species roaming the Earth.

Mollusks fall into seven classes, three of which are very large and diverse today: Gastropoda (snails and slugs), Pelecypoda (clams and oysters), and Cephalopoda (squids and octopuses). Monoplacophorans, represented by the animal *Neopilina*, are considered to be living fossils because of the simplicity of their structure. Aplacophorans are deep-sea, worm-like forms; Scaphopods are marine mollusks called tooth shells because of their appearance; and Polyplacophorans are chitons, mollusks which cling tightly to rocks in intertidal areas along the seashore.

It might seem that the slug that munches your tomatoes would have little in common with an oyster busy making pearls on the bottom of the ocean, but all mollusks share a collection of traits that reveals their common origins. Shells fashioned from calcium and protein are a major success story for mollusks, yet this feature has been lost from time to time, especially among gastropods like garden slugs and some marine relatives called sea slugs. Shells are produced by a layer of epithelial cells called the mantle, another

Variations on a Successful Molluscan Theme

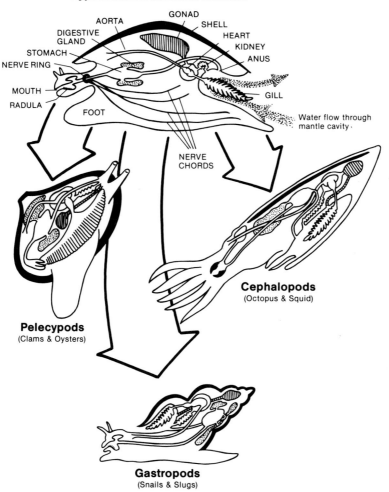

Hypothetical Ancestral Mollusk

AORTA
GONAD
SHELL
DIGESTIVE GLAND
HEART
STOMACH
KIDNEY
NERVE RING
ANUS
MOUTH
RADULA
GILL
FOOT
Water flow through mantle cavity
NERVE CHORDS

Pelecypods
(Clams & Oysters)

Cephalopods
(Octopus & Squid)

Gastropods
(Snails & Slugs)

Figure 7-7: *The first mollusk is usually considered to be a shelled creature with a muscular "foot". The radula is a distinctive feeding organ for the group and a cavity formed by the shell-secreting mantle encloses the gills and serves as a "dump" for excretory and sexual systems. This basic plan underwent at least seven variations, with three becoming very successful: the filter-feeding clams and oysters, the Cephalopods, with an internal shell and advanced nervous system, and the various kinds of snails, which underwent coiling and torsion. The latter process twisted the mantle to the anterior, placing this important area near the head. (Based on drawings in Boyle, 1981)*

molluscan feature. The edge of the mantle is an important sensory and muscular region that has undergone great modification from group to group.

At the posterior of most mollusks the mantle encloses a space called the mantle cavity, which typically houses gills and other sense organs. In addition, the digestive, excretory, and reproductive systems will discharge their products into this area.

Most mollusks move around via the contractions of a muscular foot, aided by secretions of a lubricating mucus that smooths the trail a bit. Coordination is affected by a nerve ring that loops around the beginning of the digestive tract.

The feeding organ is particularly distinctive for mollusks. It consists of a cartilaginous rod called the odontophore which is covered by a toothed band called the radula. The radula acts rather like a strip of sandpaper, scraping off algae and other bits of food that are carried back into the mouth. The radula varies from a strip of nearly identical teeth, to rows of specialized teeth, to a harpoon-like structure found in some tropical cone shells that can inject a deadly poison. New teeth are grown near the posterior attachment to the odontophores as old ones wear away at the anterior end.

Most mollusks have separate sexes, and fertilization is external. Land mollusks, however, faced with the problems of drying out, have adopted other techniques. Like earthworms, many are hermaphrodites, with both sexes present in one individual. Cross fertilization usually occurs, but some snails can fertilize

The Molluscan Mouth

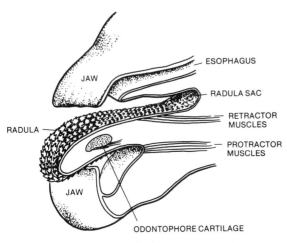

Figure 7-8: The radula is a feeding structure unique to mollusks. It is a broad ribbon containing many tooth-like structures called denticles. It is wrapped over a rod of cartilage called the odontophore that is extended during feeding. Muscles broaden and flatten the radula and erect each row of denticles in turn as they scrape over a surface. As old denticles wear out in the front, new ones are produced at the posterior end of the ribbon. There are many variations in denticle size and pattern depending on the feeding habits of individual mollusks. The cone shell has the radula reduced to a single, harpoon-like tooth that injects a deadly poison into its prey.

themselves if they turn out to be pioneers in some new environment. Land gastropods lay eggs that develop without producing the aquatic larval stage. Like annelids, aquatic mollusks produce a trochophore-type larvae that feeds on plankton. Unlike annelids, these larvae develop into a second stage called the veliger larva, which may also spend a long time as a plankton-eating organism. By eliminating these stages, terrestrial gastropods have taken the same path as reptiles in reducing their dependency on water. A few kinds have even taken the path of mammals and bear their young live.

Slugs have penises for transferring sperm across the dry gulf between individuals. Squid have a specialized arm for transferring a packet of sperm from male to female, not too unlike the solution spiders have worked out on land. Some sea slugs have solved the problem of finding a mate of the right sex by being able to reverse their sex several times in a season.

The Naked Snails

The rigors of life on land have forced slugs to develop some interesting behavior patterns. Not only do they have elaborate courtship rituals, but they also engage in territorial squabbles, particularly during the dry season.

Slugs are gastropod mollusks belonging to the sub-class Pulmonata, a word derived from the Latin word "pulmo" for lung. The mantle cavity in these creatures

The Larval Connection

Trochophore Larva
Veliger Larva of a Gastropod Mollusk

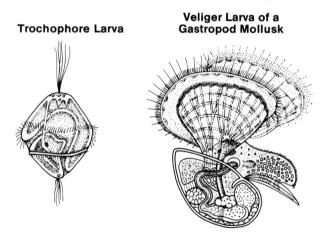

Figure 7-9: An ancient relationship between annelids and mollusks seems to be implied by the existence of very similar trochophore larvae in aquatic members of both groups. Other characteristics seem to tie mollusks to certain flatworm ancestors. Mollusks continue development into veliger larvae. Both kinds of larvae are members of the plankton, like various protozoans and small crustaceans. (Redrawn from Margulis and Schwartz, 1982)

Limax maximus
(The Giant Garden Slug)

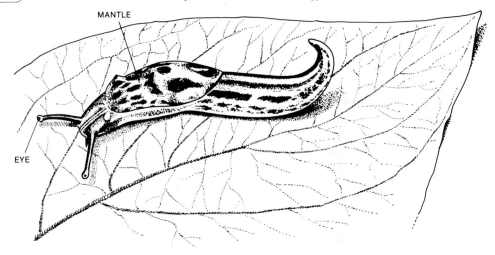

MANTLE

EYE

Figure 7-10: Limax maximus, *the giant garden slug, is a good example of a "naked snail." While the shell is gone the mantle is visible as a "hump" on its back. Garden slugs are territorial in summer months when burrows are in demand.*

has been converted into a chamber for extracting oxygen from the air. The term "slug" is a common name for land snails with reduced or absent shells. This condition may occur for several reasons. Calcium is often a limiting factor in terrestrial environments. In general, land snails have thinner shells than their marine counterparts, and if you look at many pulmonates, you can see a progression of forms with thinner and thinner shells. In some, the mantle grows over the remaining fragment of shell and you're left with a slug. Slugs gain some advantages by possessing flexible, compressible bodies that can get into small places.

Of course, a shell offers protection and a way of conserving moisture, so these advantages must be outweighed by the disadvantages—at least for this group. Slugs, it turns out, can put up with great fluctuations in moisture content. A species of the garden slug *Limax*, for example, was found to lose 2.4 percent of its initial weight per hour, or 58 percent in 24 hours (see Morton). When actively crawling, it lost 16 percent per hour through evaporation and mucus secretion. Slugs also restrict most activity to the evening, which conserves moisture.

Gastropods in general are talented at shutting down activity during cold and drought. A classic example is the species of Egyptian desert snail that spent four years as a museum specimen before walking off one day when the humidity got high!

Gastropods are also unique among mollusks in that they are coiled up and their insides are twisted around. Coiling occurs during development and results in the body mass being compacted into a spiral. This is complicated by a process called torsion, which results in the body being twisted counter-clockwise about a line drawn between the head and the tail. Internal organs are crossed over, and the mantle cavity ends up near the head of the animal. There may be at least two advantages to this arrangement: the snail can draw its head into the mantle for protection, and all the sensory equipment is near the head end where things are happening. Gastropods also show the greatest variation in radular structure and feeding behavior. Since both torsion and radular structure are not preserved in a fossil, it's often hard for scientists to tell ancient gastropods by their "covers."

Getting to Know a Slug

The time has come to search for slugs in your own backyard. Sometimes you'll find them during the day when you roll over a piece of rotting fruit or lift a board that's been lying flat for a while. The best strategy, however, is to go out when they're foraging at night. Since they're not swift and nimble, they shouldn't get away once you see them, but their drab color may make them hard to spot. Don't forget to check trees. Slugs will climb trees and plants and let themselves down on a string of slime.

Slugs rely largely on a sense of smell to find their own way around. They can find their tunnel up to three feet away, perhaps by the odor of droppings and a special "resting slime" they excrete. Slugs also leave homing trails which they can follow back to their burrows. If they stray too far away, they may follow another slug's trail but will have to contest ownership of the tunnel if the occupant is home.

Put a slug on a clear glass or plastic surface. You can see the slime trail it leaves behind. Turn the glass over and look at the muscular foot under a hand lens. Notice the ripple of contraction from back to front. Does the whole foot ripple or just a part? In many slugs only a strip in the center third of the foot is used in locomotion.

Observe the tentacles at the front end. Notice that the eyes are black spots at the tip. This is characteristic of land pulmonates. Aquatic pulmonates have the eyes at the base of the tentacles.

Do you notice a hump near the middle

Figure 7-11: *Several features of "naked snails" are visible in this photo. Note the opening to the mantle cavity or "lung" which is prominent as a distinct hole on the animal's right side. The outlines of the shell remnant are also distinct beneath the mantle. The dark spot at the tip of the tentacle in the foreground is an eye. Portions of a slime trail are evident to the right of the head and in the foreground.*

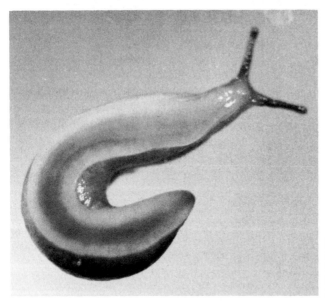

Figure 7-12: *This is a ventral ("belly") view of a slug on a piece of glass. The darker area in the middle third of the body-long "foot" ripples in waves, moving the slug to its various destinations.*

of the animal? This is all that remains of the shell sported by other gastropods.

Limax is a common genus of slugs that are relatively large and basically brown in color, mottled with dark spots. *Arion* is a genus whose members tend to be smaller than *Limax* and black in color. *Deroceras reticulatum* is a common gray garden slug. If you can collect several individuals of the same or different species you can begin to observe interactions between them.

Courtship rituals among slugs and snails have been observed for some time. Often, individuals will circle each other for long periods and at various stages will demonstrate ritualized lunging, biting, and side-swiping with their tails while waving their large penises in the air. Some species shoot chitinous "recognition darts" into their partners. *Limax maximus* climbs a tree with a partner, from which they launch themselves and mate in mid-air, intertwined and suspended by a thread of mucus.

Slug Warfare

Non-sexual aggression is apparently quite an important aspect of slug behavior—especially at certain seasons. Adversaries usually meet head to head near food or their burrows. The aggressor will touch the other slug with its tentacles. The tentacles are then rapidly withdrawn and the aggressor will touch the second slug with its mouth. The aggressor may then draw its head under its mantle, lift the forepart of its body off the ground, and lunge forward, slashing downward with its open mouth. The victim may respond by side-swiping the aggressor with its tail and quickly (for a snail) moving away. As a delaying tactic, the victim may also release a large puddle of mucus. The poor-sighted aggressor may be confused by the smell and attack the mucus first.

Alternatively, the victim may try to "tough it out" by just withdrawing its head beneath its mantle. Some slugs will slap their front ends on the ground, making them broader and flatter to try to impress an adversary with their size. The gray garden slug secretes a milky slime repellent at the site of a wound. The aggressor is put off by the taste when he gets a mouthful. *Arion ater*, the black slug, contracts into a hemispherical shape, pulls its head under the mantle, humps its back, presses its foot firmly to the ground, and rocks back and forth. If that doesn't work, it runs. A few species shed their tails, much like certain lizards. Of course, if flight and defensive tactics fail, the victim may turn and take a stand. You can find many slugs with "dueling scars" on their flanks from the abrasions of a foreign slug's radula.

Try to observe interactions between slugs and insects, also. Slugs will use much the same tactics against these six-legged foes. Often the copious slime excreted by slugs will either trip up an insect or clog their spiracles (breathing tubes) and drown them. People sometimes turn this defense against slug pests by spreading ashes in a garden. Slugs will produce so much slime in response that they dehydrate and die.

Take note of the time of year you observe aggressive behavior—especially slug-to-slug interactions. Researchers have found the behavior to be seasonal in British Columbia, with most aggressive encounters occurring in the dry months of July and August. Homing behavior is also more evident at this time. Juvenile slugs are often driven out of home burrows at the end of the season to fend for themselves. In midwinter, on the other hand, large groups of slugs may find shelter together, their differences forgotten in the common effort to survive.

If you find a good source of slugs, studying their territorial behavior could make a rewarding project. Even the most active slug has a range only a few yards square, so you won't have to crash through the back country following the beep of a radio transmitter like those who study grizzly bears or wolves. The drama is equally as great—just on a smaller scale.

The pieces of the "Cambrian Explosion" are all about you, represented largely by a host of creatures that have explored either the possibilities of segmentation and duplication of organ systems, or the various options of the mantle-secreted shell. I hope you'll seek out many of them on your own. In succeeding chapters we'll pick up a few more of the pieces by looking at various arthropod groups that combined segmentation with their own brand of external shell, and by looking at chordates, those animals who combined segmentation with an internal structural support—the bony skeleton.

REFERENCES

Boyle, P.R. 1981. *Molluscs and Man.* London: Edward Arnold Ltd.

Briggs, Derek E.G. and Harry B. Whittington. 1985. Terror of the Trilobites. *Natural History*, vol. 94, no. 12 (December). Describes the discovery of a large Burgess shale predator whose parts were once described as several different, smaller organisms.

Gaddie, Ronald E., Sr. and Donald E. Douglas. 1977. *Earthworms for Ecology and Profit*, Vol. II: Earthworms and the Ecology. Ontario, California: Bookworm Publishing Co. This is one of several references that deal with raising worms commercially.

Headstrom, Richard. 1968. *Nature in Miniature*. New York: Alfred A. Knopf. Examines the natural world by looking at what you will find on a seasonal basis.

Holder, Charles Frederick. 1892. *Charles Darwin, His Life and Work*. New York: G.P. Putnam's Sons.

Lewin, Roger. 1982. *Thread of Life: The Smithsonian Looks at Evolution*. Washington, D.C.: Smithsonian Books.

Margulis, Lynn and Karlene V. Schwartz. 1982. *Five Kingdoms: An Illustrated Guide to the Phyla of Life on Earth*. San Francisco: W.H. Freeman and Company.

McKerrow, W.S., ed. 1978. *The Ecology of Fossils, An Illustrated Guide*. Cambridge, Massachusetts: The MIT Press. A good attempt at recreating entire ecosystems from fossils.

Morholt, Brandwein and Joseph. 1966. *A Sourcebook for the Biological Sciences*, 2nd ed. New York: Harcourt Brace Jovanovich.

Morris, Simon Conway. 1979. The Burgess Shale (Middle Cambrian) Fauna. *Annual Review of Ecology and Systematics*, vol. 10, pp. 327–349. In this and the following technical reference, the authors attempt to reconstruct Burgess shale animals and speculate on their phylogenetic affinities.

Morris, Simon Conway and H.B. Whittington. 1979. The Animals of the Burgess Shale. *Scientific American*, vol. 241, no. 1 (July).

Morton, J.E. 1967. *Molluscs*. London: Hutchinson University Library.

Purchon, R.D. 1977. *The Biology of the Mollusca*, 2nd ed. Oxford, England: Pergamon Press.

Rollo, C. David and Wm. G. Wellington. 1977. Why Slugs Squabble. *Natural History*, vol. 86, no. 9 (November).

Simpson, George Gaylord. 1983. *Fossils and the History of Life*. New York: Freeman and Company.

Yochelson, Ellis L. 1978. An Alternative Approach to the Interpretation of the Phylogeny of Ancient Mollusks. *Malacologia*, vol. 17, no. 2, pp. 165–191.

8

Arthropod Invaders

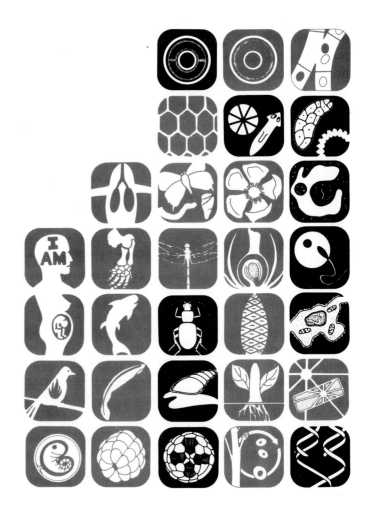

Imagine a time when the Earth was much younger, covered by large shallow oceans filled with life. A school of primitive jawless fish is feeding on a coral reef. In the distance a bed of sea lilies, something similar to starfish on stalks, is swaying in the ocean current. On the ocean floor, crab-like forms scuttle about, raising plumes of sand as they search for food. Small crustaceans, protozoans, and microscopic plants, suspended in the water, glitter like stars in the light that filters down from above.

Suddenly, a monstrous nine-foot shadow blocks out the sunlight and a fearsome claw snaps shut on one fish while the rest dart away to safety. The "monster" is a creature called *Eurypterus*, one of the largest arthropods that ever lived. He and many similar forms, collectively called eurypterids, dominated the world of their day. Most were not nearly as big as *Eurypterus*, averaging 10 to 25 cm (4 to 10 inches) in length, but existed in a host of forms. The closest modern-day relatives to these beasts are the scorpions. Scorpions, eurypterids, and whatever common ancestor they may have had, lived in a very competitive world where only the strongest or the quickest survived. Life was apparently so hard for some of these animals that they were forced to try and live where no other creature dared: the

Figure 8-1: The first animals to invade dry land appear to have been arthropods. Some of the successful ones are shown here. The scorpion in the center of the picture is an arachnid that has changed little since Silurian times. The three woodlice on the left represent the crustacean line of arthropods. The centipede on the right belongs to a group called the myriapods that may be ancestral to the insects. The head of a woodlouse (left) and millipede (right) look toward you above the Silurian-age representation of the Earth.

barren, dry, oxygen-poor surface of the world that lay beyond the mirrored ceiling of the ocean.

Figure 8-2: *Eurypterids were relatives of scorpions and spiders that terrorized the Silurian and Devonian oceans. Pterygotus, the eurypterid shown here, was one of the largest arthropods that ever lived, measuring 6.5 to 8 feet in length. The diverse eurypterid fauna 400 million years ago may have included individuals adapted to shallow waters.*

It's hard to imagine the land we walk upon as ever having been totally without life, yet for hundreds of millions of years, even as *Eurypterus* and his contempo- raries were exploiting the resources of the oceans, the land area of Earth was utter- ly barren. The Old Red Sandstones of Europe reveal a portion of the geological story. Much of the Earth seems to have been red-hued, not unlike the planet Mars today, by unreduced oxides of iron that would later be changed by the life that would invade it. There was no soil, of course, except sand, and no living sounds. The wind might howl down the course of a narrow canyon, but rustle no leaves. No insects would chirp, whir, or whine, and the sounds of birds and mammals were yet only potentialities locked in some fishy form beneath the water.

As we have already seen, plants pioneered the new environment first, but *Eurypterus'* arthropod relations were probably not far behind. In fact, the Rhynie chert of Scotland, which is some 430 million years old (lower Devonian), contains some of the earliest land plants. And tiny arachnids, relatives of your neighborhood spiders, were already making "pests" of themselves in the plants' spore-making organs. There are also remains of spider-like predators less than 3 mm long, tiny mites with piercing mouthparts, and primitive silverfish not too unlike the ones you occasionally find in your bathtub or basement. Millipede-like animals fed on the remains of these pioneer plants, and the remains of fungal hyphae testify that these colorless consumers were at work decomposing both plants and animals.

The very first invasion of the land by

Figure 8-3: *The fierce competition of the Silurian seas may have forced a scorpion-like creature to venture onto dry land. Primitive plants like Asteroxy-lon (scaly in appearance) and Rhynia (straight with knob-like sporangia on the top) were already coping with terrestrial existence.*

animals probably came several million years earlier than the creatures frozen in Scottish stone, sometime in the late Silurian. This pioneering animal, experts speculate, was probably very scorpion-like in appearance. It's clear, however, that many kinds of animals were "ready" for a land invasion. What features were important in determining which creatures would be successful?

The Arthropod Plan

The early animals that exploited dry land were all arthropods. Obviously, the arthropod body plan was somehow adaptable to terrestrial conditions. Arthropods are segmented animals—a condition apparently inherited from annelid ancestors. They have no internal skeleton, but are covered with an armor-like coating of chitin to which muscles are attached. Arthropods also possess jointed walking legs. In addition, the basic arthropod plan includes chewing mouth parts and gills that are at least partly protected by folds of chitin.

Segmentation, by itself, may not have been a large factor in preadapting arthropods to land, although compartmentalization may have helped limit loss of fluids during injury. The major groups of arthropods are distinguished by the number of body segments. Chelicerates, like spiders and scorpions, have a two-part body consisting of a head-thorax and abdomen. Mandibulate arthropods have multiple segments, with the largest class, insects,

possessing a three-segment plan of head, thorax, and abdomen. Crustaceans, millipedes, and centipedes are three other large classes of mandibulates.

The box-like external skeleton of arthropods is a significant advantage in dry-land existence. It helps keep precious moisture from evaporating, especially in those forms with a protective, waxy coating. Of course, if the chitin is too effective, oxygen can't get in and wastes can't get out, so there must be a compromise. Successful solutions all involve protecting the breathing surfaces from drying out as much as possible, because this is where most gas exchange will occur. Water-dwelling eurypterids had book gills, so called because the many leaf-like gills looked like the pages of a book. As water flowed over them, oxygen diffused in and carbon dioxide and other wastes diffused out. Land-dwelling scorpions and their relatives, the spiders, have book lungs (see Figure 9-4). These are similar to book gills in form except they are completely enclosed in the body. Air enters through a pore, the size of which can be controlled by muscles. Book lungs probably developed from book gills over the millions of years that arthropods struggled with the problems of land life.

Many arthropods, including millipedes, centipedes, and insects, solved the problem of breathing by developing tubules, called tracheae, that carry air to the body tissues or to body spaces where gas exchange occurs with the blood (see Figure 9-4). Although the system works, it constrains arthropods in a very important way: it is one of the factors that restricts the size to which they can grow. This is because there is a limit to how far oxygen can diffuse from tracheae to cells and the volume of a creature grows much faster in relation to the surface area of his skin as he grows larger.

Small size, however, does allow the arthropod's external skeleton to serve as an adequate support system outside of the buoyant water medium. Soft-bodied animals or even very large arthropods would collapse without water around them. Arthropods needed only to strengthen the musculature that already used the chitinous shell for sites of attachment. With jointed walking legs—rather than a means of movement that depended on the resistance of the water—arthropods could scuttle about on the moist sand or crawl into cracks among spray-moistened rocks.

The animals that invaded the land fed on the plants that preceded them, fellow invaders, including each other, or small creatures that lived in shallow water. They could do this because they inherited the chewing mouthparts of the fish-munching *Eurypterus* and his cousins. Animals dependent on straining their food from the water couldn't have made the crossover to dry land.

Reproduction is a difficult problem for land-based organisms because the sex cells are microscopic and dry out easily. In water, simple organisms may produce these cells in huge numbers, and currents will carry them far away. Enough will reach the proper destination to maintain the species. More complex species, like squid, for example, will produce fewer sex cells, but have developed ways of making the union of egg and sperm less chancey. Certain arms on the male are used to carry packets of spermatozoa directly into the female's body cavity. Land organisms, to become completely terrestrial, also needed precise methods to transfer sex cells between partners. Early land arthropods in the chelicerate line, like modern spiders, may have used one of their appendages called pedipalps to transfer sperm from the male to the female's gonopore. Millipedes and centipedes display several techniques, including the production of spermatophores (sperm held together in a jelly-like substance) that can be deposited and later picked up by females, and penes that can transfer sperm to holding chambers within the female's body.

An important point to remember is that the invasion of land was accomplished by many arthropod groups that possessed body forms and lifestyles that could adapt to terrestrial conditions. As the fossils from Rhynie demonstrate, animals, plants, and microorganisms, then as now, are bound together by their needs and interactions. Millipedes, centipedes, primitive wingless insect-like creatures, mites and spider-like forms, and woodlice, as well as types that have not survived to grace our backyards, all made the transition to land quite early. Millipedes ate decaying plant and animal material, centipedes ate millipedes and woodlice, spiders ate the insect types as well as each other, and . . . well, you get the idea. Life spread to the land because the land was there and offered opportunities along with all the problems.

What's a Myriapod?

In the next chapter we will look at woodlice—crustaceans that successfully invaded the land. Chapter 10 looks at spiders and Chapter 11 deals with the complex interactions of the most abundant creatures on our planet, the insects

and the higher plants. However, two other large classes of arthropods are also important land organisms. Collectively, these animals are called myriapods—literally, the many-legged ones.

Myriapods have a head and a trunk. The latter is divided into many similar segments. The myriapod head is somewhat similar to the head of crustaceans, but simpler, with fewer appendages. Myriapod legs are single, jointed legs rather than the two-branched legs of crustaceans. One branch of the crustacean leg has an attached gill. Myriapods breathe with the slender, tube-like trachea that branch throughout their tissues and open to the air at each segment. Myriapods come in two major versions: vegetarians called millipedes and carnivores called centipedes.

Millipedes (class Diplopoda) have segments which are actually pairs of fused segments. Each of these double segments, except the first three, have two pairs of legs. Millipedes have one pair of antennae. Openings from the reproductive system, the gonopores, are found on the second segment. Millipedes are typically round, although at least one variety looks similar to woodlice. In most millipede groups, the legs on the seventh segment of the male are modified as copulatory organs. The shape of these legs

Myriapods

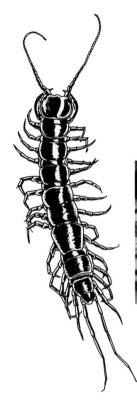

Figure 8-4: *The two major kinds of myriapods you are apt to see are centipedes and millipedes. Centipedes like Lithobius (left and in photo) are swift predators with large jaws and poison glands. They have one pair of legs per segment. Millipedes like Pachydesmus (right) are slow-moving vegetarians with two pair of legs per segment. Lithobius is about 1.2 inches long, red, and is often found under boards or stones. Pachydesmus is 2.8 inches long and lives in the southeastern states.*

is used in identification.

Centipedes (class Chilopoda) are swift predators. They tend to have flat bodies with long, seven-segmented legs—one pair to a segment. They have long antennae with 12 or more segments, and mandibles. The first pair of legs are modified into fangs with poison glands. The bites of centipedes can be painful, especially that of *Scolopendra*, but are usually not dangerous. The centipedes with the fewest legs tend to be the fastest.

Finding Myriapods

Look for millipedes under stones, in moist soil and leaf litter. Also try woodpiles and under flat pieces of board that haven't been overturned in a while. Millipedes avoid light and tend to be secretive. Many millipedes have pores along their sides that secrete materials that are poisonous or repellent to a variety of creatures. Others use a passive defense of rolling up into a ball, letting their armor discourage an enemy. Nevertheless, birds, amphibians, and other creatures find them tasty.

Females lay their eggs in soil. Sometimes they are encapsulated in egg cases. The young are born with three pairs of legs and add legs at each molt.

Centipedes tend to be found in similar habitats. You may be surprised at how quickly they can move. Some centipedes are eyeless, and many others see poorly. They appear to locate prey by smell and touch. Some centipedes produce silk which is used in prey capture, in mating rituals, or both. When a male finds a female they will touch antennae. The male may then deposit sperm in a webwork of silk which is picked up by the female. Young centipedes usually hatch out with a full complement of legs. Successive molts result in an increase in size.

Stone centipedes (Lithobiomorpha) are common in my backyard. They are fairly small, less than 4.5 cm. Adults have 18 body segments. They have 15 pairs of legs and 20 to 50 or more segments in the antennae. When disturbed, they move the last pair of legs rapidly, throwing off droplets of sticky material designed to discourage predators.

•The symphylans are a small class of myriapods you might confuse with centipedes. They are 2 to 8 mm long with 12 pairs of walking legs and 15 to 22 back plates. Unlike millipedes and centipedes, all the legs on one side move together. Occasionally they become garden pests.

Peaceful Coexistence

Arthropods account for 75 percent of all animals on our planet, and insects account for 70 percent of all arthropods. For every man, woman, and child of *Homo sapiens* in the world, there are about one million insects. It's no wonder that arthropods trouble us now and then, but we couldn't get along without them. They preceded us onto the land and forged relationships with the plant world that make them indispensible for the reproduction of many plants. Try to appreciate their success and the beauty of all their myriad forms as you discover them in your gardens.

Crustaceans, as a group, did not do well as dry-land invaders, except perhaps for the woodlice. These small arthropods, often called pill bugs, roly-poly bugs, or sow bugs, have found ways to cope with terrestrial problems with a minimum of adaptations. They are, in fact, still struggling with the rigors of land life. They may give us a clue, in the next chapter, of how primitive arthropods first dealt with the same problems 400 million years ago.

REFERENCES

Clarke, Kenneth U. 1973. *The Biology of Arthropoda*. London: William Clowes and Sons, Ltd.

Cloudsley-Thompson, J.L. 1968. *Spiders, Scorpions, Centipedes and Mites*. Oxford: Pergamon Press Ltd.

Horn, David J. 1976. *Biology of Insects*. Philadelphia: W.B. Saunders Company.

Levi, Herbert W. and Lorna R. 1968. *Spiders and Their Kin*. New York: Golden Press. Provides a simple key to the major groups of myriapods, particularly the common millipedes and centipedes. It also describes the major kinds of woodlice.

Lewis, J.G.E. 1981. *The Biology of Centipedes*. New York: Cambridge University Press. Provides a more detailed technical reference.

Meglitsch, Paul A. 1967. *Invertebrate Zoology*. New York: Oxford University Press. An excellent college text for the invertebrates in general. The older edition listed here has more information than newer, condensed versions.

Savory, Theodore. 1977. *Arachnida*, 2nd ed. New York: Academic Press.

9

Pill Bug Strategies

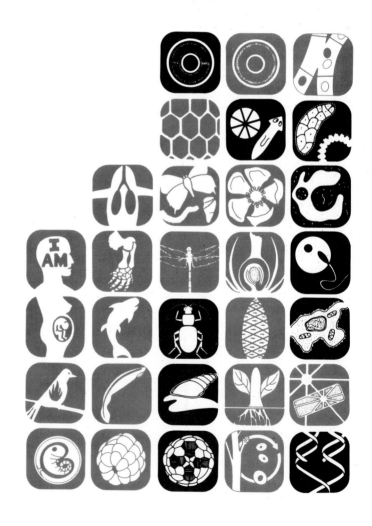

Although scorpions and their kin may have been the first animals to conquer dry land some 400 million years ago, and insects may have done the job more thoroughly, another group of arthropods tackled the problem quite successfully: pill bugs.

These unassuming little creatures are usually slate gray, sometimes with yellow spots, about half an inch long, and look very much like miniature armadillos. Like an armadillo, they roll up into a ball when disturbed. Chances are you played with them as a kid. They are unusual land creatures because nearly all their crustacean relatives live in water. Land crabs and a few other crustaceans have ventured ashore, of course, but they are restricted to moist habitats. Pill bugs are found from seashore to desert and nearly everywhere in between, including your garden. In fact, they find your well-trimmed lawn and carefully-tended gardens quite hospitable, and even if they munch on a tomato now and then, they usually pose no serious problem.

What's peculiar about pill bugs is that they seem undeservedly successful. They're a bit like the amateur athelete that bests a professional at his own sport. Pill bugs lack the waterproof waxy covering of other arthropods, their respiratory system is quite primitive, and they don't regulate the salt content of their blood like their spider and insect cousins. What, then, is the pill bug solution?

Pill Bug Problems

Before we talk about solutions it would be useful to consider a pill bug's problems in a little more detail. You will get a feeling for some of those problems if you culture pill bugs for a while. (See suggestions for raising pill bugs at the end of this chapter.) Pill bugs have two sets of concerns: one set related to just surviving on dry land and the other concerned with making a living. Put yourself in a pill bug's shell for a moment. Your aquatic cousins possess delicate gills that are admirably suited for extracting oxygen from the water. You are on dry land where gills would shrivel quickly or, at best, stick together in a useless bundle like wet hair. Moreover, temperature changes are slow in a large body of water. Aquatic crustaceans adjust by moving up or down in daily or seasonal rhythms that seem stately when compared to the adjustments you have to make to the rapid temperature changes around your local grass stems. Another, less obvious problem is finding a source of copper, the oxygen-carrying element in your blood. Then, of course, there are the traditional problems of fending off pred-

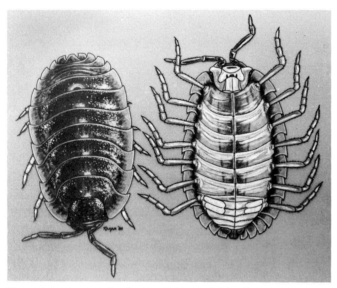

Figure 9-1: *A generalized terrestrial isopod. They have a body with seven thoracic segments each carrying a pair of walking legs and six smaller abdominal segments. The number of legs alone helps distinquish them from insects and spiders, which have three and four pair, respectively, and pill millipedes which have eleven to thirteen pair.*

flattened from top to bottom and hug the ground closely, thus efficiently distributing the pull of gravity. Isopods retain their eggs in a brood chamber. In their land-dwelling members, this helps keep them from drying out. They also possess internal fertilization—another plus for a dry land existence. Mouth parts designed for chewing rather than filter feeding also aided their land colonization. Nevertheless, water balance is a crucial factor in their lives, and it's fascinating to see how they cope.

The Water Problem

ators and parasites, finding and competing for food, locating a receptive mate, and raising a family to carry on the fight in the next generation.

Pill bugs belong to a group of crustaceans called isopods. They inherited a few things from their aquatic isopod cousins that gave them a head start in solving some terrestrial problems. Isopods possess walking legs rather than the more specialized swimming legs for getting around in a watery medium. They are

Pill bugs lose water in basically four ways: through their general body surface, evaporation from respiratory membranes, in their feces, and from nitrogen excretory organs. Studies have shown that water loss through their body surface is proportional to the vapor pressure of water in air. Also, an increase in temperature doesn't result in a sudden increase in permeability at some point. If permeability did increase rapidly at a certain temperature, it would

Armadillidium vulgare
When Rolled Up

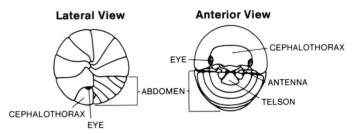

Lateral View **Anterior View**

Figure 9-2: Pill bugs roll up when disturbed and in that shape do somewhat resemble a pill. This ability helps to conserve water loss as well as provide protection from predators. (Side view redrawn from Levi's Spiders and Their Kin, front view based on an illustration in Kaestner)

imply that a water-proofing barrier had been broken down by the heat. Thus, scientists presume that pill bugs possess no water-proofing layer like the waxy coating of insects. Pill bugs may have a lipid (fat) layer beneath their outermost shell that cuts losses somewhat, but basically they are at the mercy of the ambient temperature and humidity.

Pill bugs lose some water in their feces, although the water content of their solid waste is lower than that of the food they consume. Nitrogen is excreted in the form of ammonia rather than uric acid. This results in greater water loss than for other terrestrial arthropods, but this system of waste removal is also cheaper in terms of energy expended.

Loss of water from respiratory membranes is also a serious problem. In an aquatic habitat an organism can be most efficient in extracting oxygen from water by maximizing the surface area exposed to aerated water. In small animals, simple diffusion is sufficient. In larger ones, thin, filamentous gills work admirably well. In dry air, how-

Figure 9-3: Place a number of pill bugs in a container and you will find they tend to bunch together (conglobulate). This is a moisture-conserving behavior pattern.

ever, maximizing surface area also maximizes evaporation, and thus water loss. Terrestrial invertebrates have developed several physical strategies for accomplishing this. The pill bug's solution is least efficient.

Pseudotracheal System
(Terrestrial Isopods)

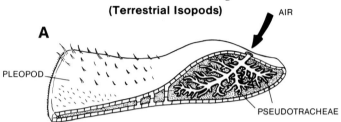

A

AIR

PLEOPOD

PSEUDOTRACHEAE

Tracheal System
(Insects)

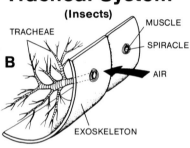

TRACHEAE

MUSCLE

SPIRACLE

B

AIR

EXOSKELETON

Book Lung
(Spiders)

C

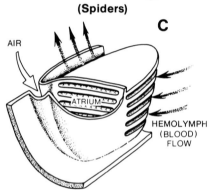

AIR

ATRIUM

HEMOLYMPH
(BLOOD)
FLOW

Gills
(Crustaceans and other aquatic arthropods)

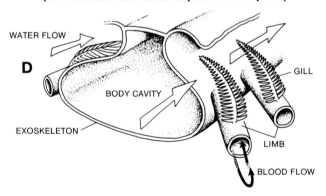

WATER FLOW

D

GILL

BODY CAVITY

EXOSKELETON

LIMB

BLOOD FLOW

*Figure 9-4: Arthropods have adopted several strategies for extracting oxygen from the air without drying out in the process. Terrestrial isopods have a workable, though somewhat inefficient, pseudotracheal system that supplements direct diffusion. (**A** is adapted from a drawing in Kaestner, **B** & **D** based on drawings from Clarke and **C** is redrawn from Foelix)*

Respiratory Strategies

As we saw in the last chapter, insects and myriapods developed a "pipe-air-to-the-tissues" approach. In this system a network of tubes called tracheae branch throughout the body and carry air directly to the tissues, where diffusion can take place. A pair of openings to the system are usually found on each abdominal segment, and the openings are surrounded by muscles that can close off the system if too much drying occurs.

Spiders and other arachnids pioneered a type of lung. This could be referred to as a "move-blood-by-a-sack-of-air" technique. In this system, air is allowed in a chamber called the atrium, and it flows from there into plate-like hollows around which blood is forced to flow. The air spaces are thin, flat and stacked one upon the other much like the pages of a book, resulting in the name "book lung." Like the insects, spiders can close the opening to the atrium if conditions are too dry.

The pill bugs' technique is similar to a book lung in some ways, but much less sophisticated. If you flip a pill bug on its back you will find two pairs of white, bean-shaped structures on the first two abdominal segments (near the rear end of the animal). These are modified portions of appendages called pleopods. The white appearance is from air trapped within hollow spaces called pseudotracheae. If you place a drop of water on the abdomen you may see a bubble which marks the opening to these air sacks. Even though they're called pseudotracheae, pill bug blood ("hemolymph") picks up oxygen as it bathes the area rather than having oxygen diffuse directly from the sack to tissues, as with insect tracheal systems. Oxygen from this source is supplemented by oxygen that diffuses directly through the cuticle. About 10 percent of a pill bug's needs are met by simple diffusion. Pseudotracheae can't be closed off by muscular action; therefore up to 42 percent of the animal's water loss is from this meshwork of air spaces.

As you will soon see, the problems involved with water regulation affect all of a pill bug's concerns, including food selection, dealing with enemies, and reproduction.

Finding the Groceries

Pill bugs will eat a variety of things, including fleshy fruits, fungi, dead and decaying matter, spider eggs, seedlings, and even ant droppings. Their food selection is largely determined by their water needs and their dependence on a source of copper.

Copper is important to pill bugs because the oxygen-carrying chemical in their blood is hemocyanin, a copper-containing molecule. There has been some disagreement among researchers, however, as to whether copper is scarce and needs to be "hoarded" by pill bugs, or if it is actually present in abundance and needs to be excreted in large quantities so that it doesn't become toxic. More work needs to be done in this aspect of pill bug ecology.

Pill bugs also eat their own feces. Feces, in fact, constitute about 9 percent of their diet. This habit, delicately called coprophagy in scientific jargon, is apparently important, because the growth rates of these animals are diminished when they are denied access to this food source. Eating feces seems to accomplish two things: it allows microorganisms to act on the material and further break down hard-to-digest items, and it provides some survival value when food is limited.

The pill bug digestive system may be more dependent on the help of microorganisms than the systems of other creatures eating much the same food. Snails and slugs, for example, have "salivary" glands, stomach, crop, and various intestinal glands to aid in the digestive process. In contrast, pill bugs have a rather basic system consisting of a tube extending from mouth to anus and two pairs of glands near the front end that contribute to digestion.

Pill bugs select food partly on the basis of its water content. This helps them to regulate the amount of salts in their body fluids. Like their marine relatives, however, pill bugs are fairly tolerant of swings in salt concentration. Insects and most other organisms are sensitive to even small shifts in the concentration of body salts. Thus, pill bugs can survive the salt problem by using behavioral strategies rather than relying on special organs to regulate water loss.

Fending Off the Foe

If pill bugs could have nightmares, a recurring one might involve images of a six-eyed demon with rapier jaws attacking suddenly out of the darkness. Dysderid spiders, seen at close range, are quite formidable looking, with specialized jaws designed to pierce pill bug armor. These spiders are not deterred by the acrid secretions of their prey, which do deter many species of spiders and other potential enemies. The secretions of pill bugs are produced by repugnatorial glands located dorsally in pairs along the margins of their thoracic segments. Although quite effective, the expense in terms of water loss may be great, for the size of these glands decreases in species that are progressively adapted to drier and drier habitats.

Another unusual defense has been adopted by a European pill bug. Its protective coloration makes it look nearly identical to the abdomen of the European black widow spider. Presumably, this offers protection against certain lizards which, after sampling a widow and feeling poorly as a result, avoid anything with the same distinctive red and black coloration.

Birds, amphibians, mites, and centipedes also take their toll on pill bugs, although a complete list of predators has probably not been compiled. For many of these predators the "active" defenses of pill bugs are not a great factor. Rather, it is the pill bugs' preference for dark, narrow burrows and sheltered areas that keeps them hidden from most of their enemies.

At least one species of parasite, an acanthocephalan worm, has managed to "reprogram" pill bug behavior to its own advantage. Janice Moore at Colorado

Figure 9-5: If pill bugs had nightmares this six-eyed Dysderid spider would probably be in them. Dysderids have specialized jaws for piercing pill bug armour and apparently are not put off by the secretions from their repugnatorial glands.

State University has followed the life cycle of this enterprising creature whose primary host is the starling. The worm reaches sexual maturity within the starling and lays its eggs in the bird's intestines. The eggs are passed with the feces, which are in turn eaten by pill bugs. In this secondary host the young worms hatch and burrow through the gut wall. They either remain attached to the gut or break away to float freely in the space between organs. There the worms grow and share the pill bug's dinner, reaching a size of 2 to 3 mm. Since pill bugs average about 8 mm long, a substantial fraction of their body cavity is occupied by the intruders. Infected female pill bugs have been found to be sterile. The most interesting aspect of the relationship, however, is that by some

unknown mechanism the presence of the parasite alters the pill bug's normal behavior in such a way that it becomes more vulnerable to the bird predators, the worm's primary host. Infected pill bugs were found more frequently in areas of low humidity and showed no preference for sheltered areas—behaviors unlike their uninfected brethren. These tendencies kept them in exposed areas where starlings were more likely to find them.

Pill bugs become most susceptible to all their enemies—and the environment—during molting periods. Females undergo extra reproductive molts, and not surprisingly, their mortality is higher than that of males. Overall mortality for both sexes is fairly high, with only about a fifth of pill bugs surviving the first six months and about one percent making it to the ripe old age of four years.

A Pill Bug Mimic

LACTRODECTUS MACTANS	ARMADILLIDIUM KLUGII
(European Black Widow)	(Pill Bug)
Ventral View	**Dorsal View, rolled up**

Figure 9-6: *A species of European pill bug has adopted the same color markings as the European black widow spider—a technique that apparently protects it from predation by lizards. (A. klugii based on drawing in Levy's paper; L. mactans redrawn from Levi's Spiders and Their Kin)*

Raising a Family

Pill bugs inherited from their aquatic relatives a structure called the brood pouch, which served as a useful preadaptation to land life. Females shed their eggs into a chamber formed by plate-like overlapping extensions of the first five thoracic legs. Within this chamber the young de-

Cross-section Through Thorax of *Oniscus*

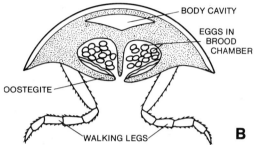

BODY CAVITY

EGGS IN BROOD CHAMBER

OOSTEGITE

WALKING LEGS

B

Figures 9-7A &7B: *The photograph shows a female* Armadillidium vulgare, *ventral side up, with a brood pouch full of eggs. The brood pouch is created by overlapping, plate-like extensions of the first five pairs of walking legs. Aquatic isopods also possess brood pouches, a situation which may have helped pre-adapt this group for a terrestrial existence. (Cross section redrawn from Lawrence)*

velop over a period from two to three months, relatively secure from environmental dangers. Brood size varies from a couple of dozen offspring to several hundred, depending on the age of the female. The older and larger the female, the more young she will carry. In California grasslands, pill bugs have two broods per year, breeding from April to September. The same species, *Armadillidium vulgare*, in England usually produces one brood per year. Gravid females can be found from May through the summer months.

Females do need males to produce fertile eggs (they're not parthenogenic like aphids—see Chapter 11). However, sperm can survive in the female for up to a year. Curiously, broods can be mixed with regard to sex of the offspring or they can be comprised of all males or all females. One possibility is that the sex chromosomes are selectively segregated during meiosis, such that the sex-determining partner (female in the case of pill bugs) contributes only one kind of sex chromosome in those individuals that produce broods of one sex.

French researchers have recently found that in some populations of *Armadillidium* there are F-factors, presumably virus-borne or loose scraps of DNA in the cytoplasm, that can transform genetic males into apparent females and genetic females into intersexed animals. The geographical extent of the F-factor, however, seems to be limited. When the young pill bugs crawl out from the protective flaps of their brood pouches they are on their own. If conditions are too dry, they will lose water quickly, and if too moist, they will succumb to fungi. Brood mortality in one study was 7 to 8 percent. Young pill bugs are nearly white and become progressively darker in successive molts, eventually turning a leaden gray or gray-black with sulfur-yellow spots.

At one year of age females are ready to breed. Although one-year females are the largest part of the breeding population, they are out-produced by two-year individuals which have larger broods.

The Pill Bug Solution

The pill bug's solution, not only to surviving on land with a rather meager collection of physical adaptations but also to outcompeting other soil inhabitants, has been to develop a simple, but effective, repertoire of behaviors. As you observe pill bugs in your own gardens, see if you can recognize some of the following activities.

Coping With
Water Loss and Surplus:

Pill bugs use nearly all their behavioral options in dealing with water economy because it is a central problem in living on land. Loss of water is a function of air temperature, relative humidity, rate of evaporation, and the nature of the soil. A pill bug's major mode of dealing with this problem, as well as others, is to move. It can move toward desirable conditions, away from undesirable ones, and move in circles if it likes things the way they are. The speed of movement reflects the nature and severity of the stimulus. A pill bug's response to light, temperature, and humidity is determined by a complex interaction between these three variables. Pill bugs' responses are also not always the same as those of other terrestrial isopods. Although Figure 9-8 summarizes some of these differences, you may want to look at some of the papers of Warburg and other researchers if you're particularly interested in this area.

A pill bug's response to light is diminished toward sunset. At high temperatures, also, movement is faster and in a straight line, whereas at moderate temperature there is more turning, which tends to keep the animals in more hospitable temperature ranges.

Pill bugs can actively take up droplets of water if they become dried out. You can observe them in a petri dish, rhythmically contracting their abdomens and drawing water up between their tail projections as if through a straw. Grooves running

ENVIRONMENT	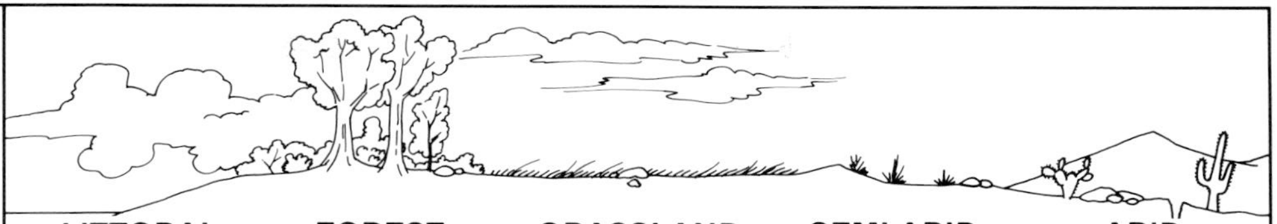 LITTORAL	FOREST	GRASSLAND	SEMI-ARID	ARID
TERRESTRIAL ISOPODS: PHYLUM: ARTHROPODA CLASS: CRUSTACEA SUBCLASS: MALACOSTRACA ORDER: ISOPODA	Family: Ligiidae 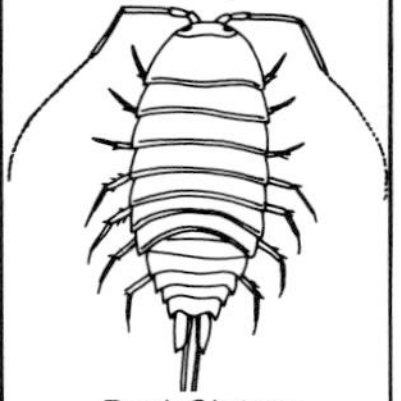 Rock Slaters	Family: Oniscidae 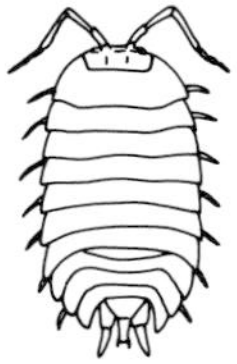 Sow Bugs	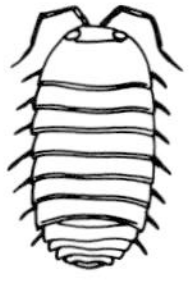 Family: Armadillidiidae Pill Bugs	Family: Porcellionidae 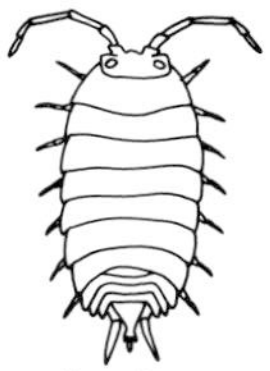 Sow Bugs	Family: Armadillidae 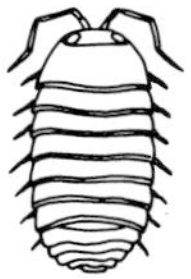 Pill Bugs
Can they roll up? (Conglobate)	No	No	Yes, excluding antennae	No	Yes, including antennae
Reaction to Light	Negative, except at high temperatures	Negative	Positive, except at low temperatures	Negative, except at high temperatures	Negative
Reaction to humidity	Positive	Positive	Positive, except at high temperatures	No data	No data
Reaction to temperature	No data	Positive	Positive	No data	No data
Water loss	High	High	Limited	Limited	Very limited

Trends

→ Reaction to light becomes more important than reaction to humidity

→ Decrease in size of tegumental glands

Figure 9-8: Terrestrial isopods are found in a full range of soil ecologies. Although their physiology and anatomy shows some specialization in the various habitats, it is behavioral adaptation that largely accounts for their success under very different conditions. (Data for the chart taken from Warburg, 1968)

from the base of one leg to another carry water the full length of the body on either side. Fanning motions of the pleopods speed the process. A drop of colored water placed on a pill bug's back soon ends up in these lateral grooves—collectively called a water transport system—and is carried toward the head where they can drink it.

Getting rid of excess water is a matter of evaporation and seems to partially dictate pill bug's nocturnal behavior, allowing them to get rid of moisture accumulated in their daytime burrows. Pill bugs quickly lose water from the ventral side of the abdomen where the respiratory organs are located. Just keeping the abdomen flat to the ground greatly reduces water losses. Thus, it is not surprising that pill bugs prefer to be in contact with something on all sides. They also tend to bunch with other pill bugs quite often, something you can readily see if you collect a few and allow them to settle down on soil or sand. A pill bug's tendency to aggregate is regulated by more than a liking for close contact, however. They are attracted to each other's odors—more so as they get drier. It seems to be this behavior that gets them back to shelter before daybreak. The behavior which gives a pill bug its name—the ability to roll up into a pill-like shape—also helps conserve water, although whether that is more important than its

antipredator function is uncertain. Other types of terrestrial isopods lack the ability.

Temperature Extremes:

Overheating, independent of any water loss involved, is a more severe problem for land organisms than for their water-based relatives. Most terrestrial isopods move away from light, and this tends to keep them in shaded, cooler places. Pill bugs are somewhat unusual in this regard because they are attracted to light except at high temperatures and humidity. Light, for pill bugs, is probably more of a general indicator of environmental conditions than a strong determinant of behavior.

As mentioned earlier, pill bugs move more quickly and in straight lines away from areas of high temperature. Evaporative cooling also lowers a pill bug's body temperature. Their nocturnal behavior makes them active in the cooler, nighttime hours. Bunching behavior most likely results in more consistent collective temperatures. A temperature between 20° and 30°C seems to be optimal for aggregation.

Pill bugs can also adapt to temperature extremes if given some time to acclimate. If *A. vulgare* are collected at low temperatures and kept two weeks at a temperature of 30°C they will survive for 30 minutes at

40°C, whereas non heat-adapted animals will die right away.

Respiration:

Respiration is an activity that is perhaps least subject to supplementation by behavioral techniques, but some terrestrial isopods do increase the flow of air over their pleopods by movements of the abdomen. Upward movement draws air in, and it enters air spaces created between the pleopods at the junction between the thicker part (that contains the pseudotracheae) and the thinner, plate-like portion. Downward movement of the abdomen expels the air. Oxygen diffuses directly across the moist membranes.

While insects may represent the pinnacle of arthropod adaptation to a terrestrial environment, and spiders can claim a close lineage to the first terrestrinauts in the animal kingdom, pill bugs and their kin have taken the basic crustacean body plan and "behaved" their way into an important part of your backyard ecology. They are excellent subjects for studying how the interplay of environmental conditions affect animal behavior. At least that can be your excuse the next time you put one on your hand and watch it roll up.

Finding and Raising Pill Bugs

How do you know a pill bug when you find it? The terrestrial isopods, of which pill bugs are one example, include rock slaters and sow bugs. Cloudsley-Thompson (1968) lists other common names—at least in England—such as bibble bugs, coffin cutters, and tiggy hogs. Rock slaters are relatively primitive and inhabit shorelines. Sow bugs are common in leaf litter and humus on forest floors, and pill bugs are populous in grasslands and deserts.

All of these terrestrial isopods share some common traits. Their basic body shape is oval with an arched back, much like a lozenge cut in half lengthwise. They have a pair of eyes on the head which are not stalked like lobsters and other decapod crustaceans. They have two large antennae and two smaller antennules in front of them. The latter, however, are hard to see without magnification.

The thorax of terrestrial isopods has seven segments which are usually broader than the six abdominal segments which follow. Each thoracic segment has a pair of walking legs. Thoracic segments 2 to 5 have ventral plates that form the brood pouch in pregnant females. The abdominal segments have appendages called pleopods. The inner branches, or rami,

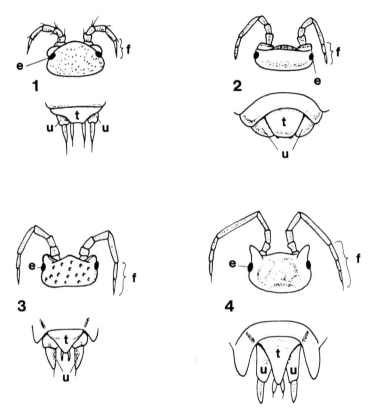

Figure 9-9: *This figure shows the anterior and posterior ends of four common woodlice. 1) Trichoniscus pusillus 2) Armidillidium vulgare 3) Porcellio scaber and 4) Oniscus asellus. e=eye; f=flagellum of antenna; u=uropod; t=telson. (Redrawn from J.L. Cloudsley-Thompson, Land Invertebrates, 1961, who credits the work of Edney, 1953)*

form a surface, while the outer parts serve as a protective covering. In some species, including pill bugs, you will find bumps (white in living animals) on some of the pleopods. These are networks of respiratory tubules called pseudotracheae.

Once you know you have a terrestrial isopod, it becomes a matter of knowing heads and tails. The antennae of terrestrial isopods differ in the number of segments in the terminal portion, or flagellum. The tail ends differ in the shape of the terminal segment, or telson, and the uropods beneath them (see Figure 9-9). In pill bugs the tail end is always rounded with a plate-like telson and uropods.

From a behavioral standpoint, if it rolls up, chances are you have a pill bug—in other words, a member of the family Armadillididae or Armadillidae. The common pill bug is *Armadillidium vulgare,* a member of the former family. If you flip it on its back you can see two pairs of white tracheal lungs on the first two abdominal segments. Members of the Armadillidae, the latter family, are found in desert regions and have five pairs of tracheal lungs. Pill bugs have a hard time righting themselves once on their back and will eventually stop moving their legs. This makes them easier to work with than some of the sow bugs that don't roll up. These latter animals move more quickly and, since they tend to be flatter, can right themselves more easily.

Culturing pill bugs is also quite simple, which is a plus for their use in the lab. Howard (1940) used petri dishes with damp filter paper on the bottom. To this he added dead leaves and twigs to fill the dish. Four to ten individuals could be kept in each one. He also used #1 jars with an inch of well-wetted cotton wool at the bottom. To this he added dead leaves, twigs, and beech wood leaf mold.

For my first colony I found a well-populated source of pill bugs in the clay soil beneath a cement drain spout culvert. With a shovel I removed the top inch or so of soil and placed it in a five-gallon aquarium and covered it with a piece of glass. From time to time I added a small amount of fleshy fruit or a few of the local vine weeds that they seemed to feed on. If the soil looked dry, I sprayed it with water. The key is to provide sufficient moisture without encouraging a lot of mold growth. Pill bugs do eat some mold spores, but young pill bugs are susceptible to mold. Moist sand can be substituted for soil and is a little cleaner. Carrots (or almost any vegetable or fruit) can be used as food and removed when it starts to mold. Add some powdered chalk to serve as a source of calcium for the pill bug's exoskeleton.

REFERENCES

Since pill bugs are neither saints nor sinners in their dealings with human beings, there are not a great many general references. I have listed a few below that are less technical than most.

Cloudsley-Thompson, J.L. 1961. *Land Invertebrates*. London: Methuen and Company, Ltd. Good black-and-white pictures of different species.

Cloudsley-Thompson, J.L. 1968. *Spiders, Scorpions, Centipedes and Mites*. Oxford: Pergamon Press Ltd.

Kaestner, Alfred. 1970. *Invertebrate Zoology Crustacea*, Vol. III. New York: Interscience Publishers.

Levi, Herbert W. and Lorna R. 1968. *Spiders and Their Kin*. New York: Golden Press. A good, inexpensive general reference which keys out common terrestrial isopods.

For those interested in looking into the technical literature, I suggest the following:

Allee, W.C. 1926. Studies in Animal Aggregation: Causes and Effects of Bunching in Land Isopods. *Journal of Experimental Zoology*, vol. 45, pp. 255–277.

Barrington, E.J.W. 1967. *Invertebrate Structure and Function*. Boston: Houghton Mifflin Co.

Clarke, Kenneth U. 1973. *The Biology of the Arthropoda*. New York: American Elsevier Publishing Company.

Dallinger, R. and W. Wieser. 1977. The Flow of Copper Through a Terrestrial Food Chain. *Oecologia* (Berl.), vol. 30, pp. 253–264.

Edney, E.B. 1968. Transition From Water to Land in Isopod Crustaceans. *American Zoologist*, vol. 8, pp. 309–326.

Foelix, Rainer F. 1982. *Biology of Spiders*. Cambridge, Massachusetts: Harvard University Press.

Hals, Gary D. and Kathleen G. Beal. 1982. Death Feint and Other Responses of the Terrestrial Isopod, Porcellio Scaber. *Ohio Journal of Science*, vol. 82, no. 2, p. 94.

Hassall, Mark and Stephen Rushton. 1982. The Role of Coprophagy in the Feeding Strategies of Terrestrial Isopods. *Oecologia* (Berl.), vol. 53, no. 3, pp. 374–381.

Howard, H.W. The Genetics of *Armadillidium vulgare* Latr. I. A General Survey of the Problems. *Journal of Genetics*, vol. 40, pp. 83–108.

Huxley, Thomas. 1878. *A Manual of the Anatomy of Invertebrated Animals*. New York: D. Appleton and Co.

Juchault, par P. and J.J. Legrand. 1981. Contribution a l'etude qualitative et quantitative des facteurs controlant le sexe dans les populations du Crustace Isopode terrestre *Armadillidium vulgare* latreille. *Arch. Zool. Exp. Gen.*, vol. 122 (May), pp. 117–131.

Kuenan, D.J. and H.P. Nooteboom. 1963. Olfactory Orientation in Some Land Isopods (Oniscoidea, Crustacea). *Entomologia Experimentalia et Applicata*, vol. 6, pp. 133–142.

Lawrence, R.F. 1953. *The Biology of the Cryptic Fauna of Forests*. Cape Town, Africa: A.A. Balkema.

Levy, H.W. 1965. An Unusual Case of Mimicry. *Evolution*, vol. 19 (June), pp. 261–262.

Moore, Jance. 1984. Parasites That Change the Behavior of Their Hosts. *Scientific American*, vol. 250, no. 5 (May).

Paris, O.H. 1963. The Ecology of *Armadillidium vulgare* (Isopoda, Oniscoidea) in California Grasslands. *Ecology*, vol. 43, pp. 229–248.

Pratt, Henry Sherring. 1935. *Common Invertebrate Animals*. Philadelphia: P. Blakeston's Sons and Co. Provides a more technical key to the different species.

Warburg, M.R. 1964. The Response of Isopods Towards Temperature, Humidity and Light. *Animal Behavior,* vol. 12, pp. 175–186.

Warburg, M.R. 1968. Behavioral Adaptations of Terrestrial Isopods. *American Zoologist,* vol. 8, pp. 545–559.

Wieser, Wolfgang. 1979. The Flow of Copper Through a Terrestrial Food Web. In *Copper in the Environment, Part I: Ecological Cycling,* J. Nriagu, ed. New York: Wiley Interscience.

10

Fearsome Hunters, Careful Dancers

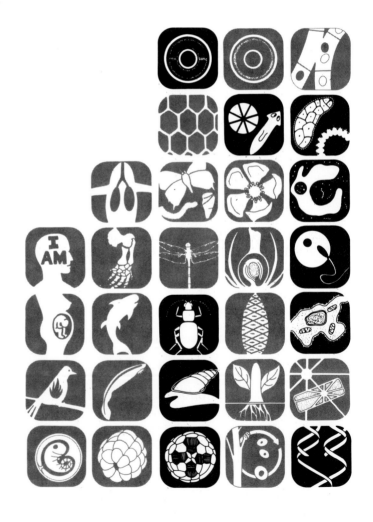

Courting or being courted, as the situation applies, is a trying experience. The world often seems to hinge on a successful relationship with the opposite sex, as perhaps it does. As difficult as courting rituals seem to be for humans, however, for some animals it is literally a matter of life or death. Many male spiders, for example, must wave their appendages in just the right manner and dance the proper dance or they could well end up on their prospective mate's luncheon menu. Such is the case with neighborhood jumping spiders such as *Phidippus*, a black and white hairy little fellow with green jaws and red fangs. But before we talk about his story in particular, it would be well to say a few things about spiders in general.

I hate to see a person react to another living creature by wanting to step on it. Nevertheless, spiders suffer this fate on a regular basis. Perhaps it's a particular aversion on the part of primates, or a general distrust of things that are predatory, or perhaps just an informal feeling that spiders are especially "icky." Spiders don't deserve such a fate, however, and do, in fact, do everyone a favor by keeping insects and other arthropods in check. Once you get beyond their intimidating appearance, they have a beauty and variety that is quite staggering.

Some 30,000 species of spiders have already been named, but this is a figure which experts feel may represent only one fourth of the total. Since experts are traditionally conservative, don't be surprised if many more eventually turn up. A much larger number of species than you would ever imagine are right there in your backyard.

Go outside on a sunny day and scan along the siding of your house. If you see a rather compact, dark spider spreadeagled there, walk over to it. Did it move with

Figure 10-1: *The orchard spider,* Leucauge mabelae, *appears to have trapped the moon in her web. Actually,* L. mabelae *is a beautifully colored spider about ⅜ inch long.*

Figures 10-2A & 2B: *The jumping spider,* Phidippus, *looks at the camera* **(A)** *while dining on an insect. The pair of primary eyes and one pair of secondary eyes just behind and to one side of the primaries are easy to see. One lateral eye is visible. Bright green chelicerae hold the insect while fangs inject digestive enzymes.*
    Phidippus *wins a confrontation with predatory wolf spider in* **B**.

Figure 10-3: *One of the few poisonous species of spider is the black widow,* Latrodectus mactans. *In this photo you can see the typical hourglass shape on the abdomen of the female (males are smaller and don't bite). The widow's round, black, shiny body, long legs and irregular web make it fairly easy to identify. Young widows may have red blotches on their back side as well, along with thin white stripes laterally. Widows are usually shy, but are apt to be seen near new construction or disturbed habitat.*

quick, jerky motions as you approached and turn to face you? If so, it's probably a jumping spider, a very keen-eyed hunter who uses his eight eyes to good advantage. Or perhaps you saw a large, yellow and brown spider suspended in its web near a window well. This is a common "orb-weaving" spider that waits for lunch to come to it as it sits, with some majesty, in the center of its world.

Search among the grass stems. Pale yellow and gray *Tibellus* stretches out there, trying to disguise itself from potential prey. Crab spiders sit on leaves and flowers, often nearly invisible against their background, and extend their first two pair of long legs to the side, waiting with "open

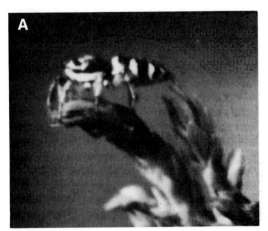

Figures 10-4A & 4B: Jumping spiders like this striped Salticus scenicus *will crawl to a projecting point and may wave their front pair of legs before making a leap.*

arms" for a careless insect. More conspicuous brown wolf spiders hunt there, too. Yellow and white body markings and long legs make them easy to see, and they will move quickly out of your way. "Daddy long-legs" (which are not actually spiders, but near relatives, Opilionida) are most likely found in your garden. Inside, you probably share living space from time to time with the common house spider.

Recognizing a Spider

How did you know a spider when you found it? As arthropods go, they're fairly easy to recognize. Their body is divided into two major parts, the cephalothorax (head-trunk) and abdomen, which is connected by a thin stalk called the pedicel. They have eight legs instead of an insect's six, and they have no antennae or wings. As you'll see when you take a closer look, their mouth parts are also different than those of insects and, of course, they make strands of sticky silk for prey capture, house construction, drag lines and, occasionally, parachute manufacturing. Spiders usually have eight simple eyes whose arrangement is often the key to distinguishing the spider families. Insects and many other arthropods have a pair of compound eyes. Although these compound eyes are structurally more complex than spider eyes, some spiders, particu-

The Zebra Jumper

Figure 10-6: Orb weaving spiders sit in their webs and wait for a meal to blunder by. Many are brightly colored like this member of the genus Araneus.

Figure 10-5: Salticus scenicus, *usually referred to as the zebra spider, is a small jumping spider found commonly on the sides of houses. This one poses on a sprig of juniper.*

larly the Salticids (jumping spiders), have very acute vision.

Internally, the cephalothroax contains the brain, poison glands, and stomach. The latter organ is called a pumping stomach because it vacuums up the remains of prey that have been predigested outside the spider's body by enzymes egested from the oral cavity through the fangs. The abdomen contains heart, digestive tract, book lungs and other respiratory organs, and silk glands. Silk is re-

leased through tiny spigots located at the rear of the abdomen on the tips of fleshy organs called spinnerets.

If you flip a spider on its back, the spinnerets are usually easy to see as a cluster of tube-like projections at the very end of the abdomen. Just anterior to them you may see a projection called the colulus, whose precise function is not known. In other spiders you will see a flattened plate called the cribellum that is used, in conjunction with special leg hairs, to add a woolly texture to spun silk that apparently helps in prey capture. Move to the anterior end of the abdomen and you can see (at least in living spiders) two lighter patches on either side of the midline that mark the location of the book lungs. They open to the outside through a pair of slits. Just posterior to them is a hard

Figure 10-7: This crab spider sits with open "arms" on the leaf of a russian olive tree awaiting an unwary insect.

plate (in most female spiders) called the epigyne, which partially conceals a slit beneath it, the gonopore, through which eggs pass.

The cephalothorax, or head-trunk region, is the attachment point for all the walking legs, mouth parts, and the pedipalps. The pedipalps are used to help handle food, and as a sexual organ in males. Males pick up sperm with their pedipalps from a web on which they have deposited a sperm droplet. The end of the pedipalp serves as an effective sperm storage receptacle which they can use to transfer sperm to a willing female to be stored in her epigynum until she is ready to lay eggs.

Surveying the Neighborhood

If you took my suggestion of looking along the house siding and in the grass and nearby bushes, you've seen a few of the spiders that cohabit with you. For a more exhaustive survey try these techniques for collecting:

The pit trap is particularly good for finding ground spiders. Dig a hole large enough to bury a tin can or bottle to its rim. Put a little antifreeze (ethylene glycol) in the bottom of the container. Keep rain and dirt out by covering with a small board that is supported by rocks underneath. Check your trap on a regular basis. WARNING: Pets and children are attracted to the sweet taste of antifreeze, which is poisonous.

A sweep net used for catching insects will also get you a nice selection of spiders if you sweep shrubs and weeds. Burrowing spiders may have to be dug out. The best way to do this is to place a thin stick or straw down the burrow, then dig down one side of the burrow following the stick to the bottom.

Hunt for night spiders with a **headlamp** or **flashlight.** Wolf spiders' eyes will glow back at you. You will also find some web spiders that come out at night.

Pile leaf litter into a **tullgren funnel,** which is a large funnel with a quarter-inch wire screen in the bottom. Support the funnel so that you can put a container with water or alcohol beneath it. Put a board with a mothball suspended beneath it over the funnel. This will drive any animals

Phidippus Anatomy

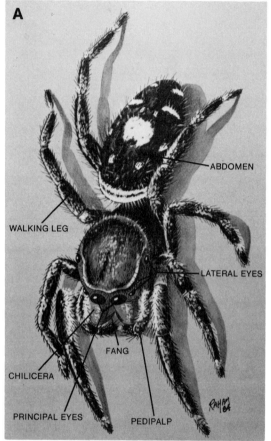

A

ABDOMEN

WALKING LEG

LATERAL EYES

FANG

CHILICERA

PRINCIPAL EYES

PEDIPALP

Ventral (belly) side of a female *Phidippus*

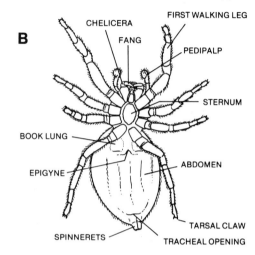

B

CHELICERA

FANG

FIRST WALKING LEG

PEDIPALP

STERNUM

BOOK LUNG

ABDOMEN

EPIGYNE

TARSAL CLAW

SPINNERETS

TRACHEAL OPENING

Figures 10-8A & 8B: The anatomy of Phidippus is similar to other spiders except for the large principal eyes and the extra-long first pair of walking legs. Pedipalps on some species are large and "showy," but shouldn't be confused with legs. The epigyne is a hard plate that is part of the duct system for receiving sperm from the male. The epigyne is just in front of the gonopore, through which eggs are released. In some spiders the females have no epigyne and males place sperm directly into the gonopore.

from the leaf litter. As an alternate to the mothball, suspend an incandescent light over the top of the funnel. The heat will also drive the animals to the bottom.

The **child and bottle** technique also works very well. Give some kids a few bottles, and let them know you want spiders. You will soon have a large selection. WARNING: All spiders are poisonous, some seriously so. Care must be taken when they are captured in bottles to prevent being bitten.

Spiders are soft-bodied and have to be preserved in liquid rather than pinned. A solution of 70 percent rubbing alcohol will work well. You can add enough glycerin to make up 5 percent of the total to keep the animals more flexible. To make a collection useful, location labels should be made with good quality paper and India ink. The labels should be put inside the vials, as exterior labels can be lost or smudged. One spider should be put in a stoppered vial, but a group of vials can be kept in a larger container.

Living spiders can be kept healthy in captivity for quite a while. This gives you the opportunity to see their true colors and behavior close up. Petri dishes, if available, work well because they are easy to see through and manipulate under a dissecting microscope. However, you can also keep spiders in small plastic boxes that you have cut a hole in with a hot knife. Tape wire mesh over the hole. Spiders do well in regular terraria; web weavers can be kept in wooden frames with glass, acetate, or cellophane sides. Small containers have some advantages. You can keep more kinds of spiders in a smaller area, of course, and you can put the containers in the refrigerator or freezer for a few minutes if you want to slow down a speedy spider for closer examination.

Spiders eat a variety of insects, millipedes, mealworms, woodlice, and other spiders. You should start a colony of fruit flies (see box on page 191) because spiders eat these readily. (Lizards, frogs, and other creatures also find them tasty.) Spiders do need access to small amounts of water, which can be provided by moistening a small cotton ball.

Jumping Spiders

Now that you've looked over the neighborhood a bit, let's get back to the problems of a *Phidippus* male spider again. As you may recall, he is the black and white jumping spider that finds himself in mortal danger if he doesn't approach his girl friends in just the proper way.

Jumping spiders belong to the family Salticidae and thus are often referred to as salticids. Older references may put them

A Tullgren Funnel

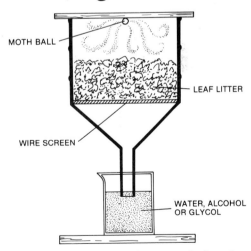

MOTH BALL

LEAF LITTER

WIRE SCREEN

WATER, ALCOHOL OR GLYCOL

Figure 10-9: Tullgren funnels consist of a funnel with a 1 cm (¼ inch) screen at the bottom. Leaf litter is placed on the screen and the fumes from a mothball force creatures into a jar of water or alcohol. A variation on this theme uses a 100 watt light bulb instead of the mothball.

in the family Attidae. Whatever name they are given, however, they are easy to recognize because of their "nervous" behavior. They move quickly when you approach, change directions easily, and will probably hop or jump to try and get away.[1] If you look at them closely you will see that the first or third pair of legs are quite a bit longer than the others. Also, the head-trunk (cephalothorax) is often larger than the abdomen in *Phidippus* males, which gives them somewhat of a "gorilla" look. If you look at them very closely (for which you would probably need a hand lens or dissecting scope), you can see they have two large eyes in front, with two others just to the side, and yet another two pairs on their "sides" (mid-lateral cephalothorax). Jumping spider jaws are often brightly colored. In *Phidippus* they are a bright, iridescent green, tipped with red fangs that fold underneath.

Phidippus or his relatives can be found in many terrestrial habitats. Jumping spiders make their homes in settings that range from intertidal areas to 6,000 m above sea level on Mt. Everest, and from deserts to rain forests and oceanic islands. Their 4,000 or so species live on every continent except Antarctica. So consider

your backyard graced by a widespread, successful family!

Jumping spiders owe much of their success to something they share with primates like you and me. They have great eyes. Their smaller eyes are good at detecting motion and give them a wide visual field, while the two big eyes in front form clear, camera-type images at close range. Experiments have indicated that they probably have color vision, too, although they can't see in red light.

Phidippus and other jumping spiders use their good eyesight during courting, although they do seem to have "backup" chemical and sound signals that are important. To see some spider interactions, capture a *Phidippus* or two. I've found that the best way is to just put a bottle over them. Get them on a flat space where they can't retreat into a crack or under a rock. Then transfer them to a small clear container or make yourself a "spider arena." I constructed one from a piece of cardboard, some heavy acetate, tape, and a piece of picture glass (see Figure 10-10). The acetate can be purchased from an art supply store. It comes in rolls or sheets of different thicknesses. Get a small amount of the heaviest quality so it won't be crushed when you put the glass on top. It would also be useful to have a small mirror (like you get at the drug store or from an old compact) and a little bit of clay to form

---

[1]Robert R. Jackson describes an Australian jumping spider which is atypical. It builds webs and is camouflaged to look like a dead leaf. It waits patiently for prey or plucks at other spiders' webs to draw them within striking range.

A Spider Arena

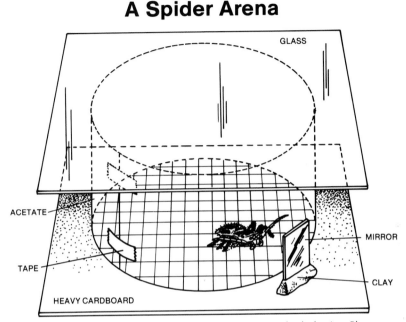

Figure 10-10: *Construct a spider arena for observing spider behavior. Place a centimeter grid in the bottom to estimate sizes and distances. (See text for details.)*

a base for it. If you line the cardboard with a gridwork of one-centimeter or half-inch lines, you can estimate spider sizes and distances quite well.

Jumping

One of the first things you will notice when you get a *Phidippus* or other salticid in your arena is its ability to jump. It may jump only a couple of centimeters when stalking prey, but can jump up to 25 times its body length. This isn't as good as your average grasshopper, but *Phidippus* and his relatives don't have specialized jump-

ing legs either. You'll notice that he uses the third or fourth pair of legs, not the long first pair.

You'll also notice that *Phidippus* doesn't go anywhere without his safety line. In a short time his container will be criss-crossed with strands of silk anchored by small white spots.

The Hunt

Prey capture is interesting to watch. Fruit flies, other small insects, or smaller jumping spiders are stalked and eaten. If you have your spider in an arena with a

A Jumping Spider's Jump

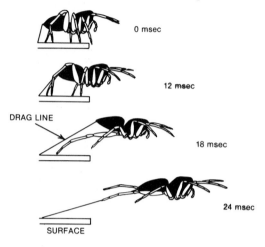

0 msec

12 msec

DRAG LINE

18 msec

24 msec

SURFACE

Figure 10-11: A jumping spider's long first pair of legs are not used in a jump, but rather the fourth pair. They affix a drag line before they leap. (Redrawn from Foelix, 1982)

entirely by the chelicerae. Venom is injected through the fangs into the prey. The digestive fluids are forced out the mouth. Spiders' jaws are not adapted for chewing, so the digestive enzymes predigest their meal. The sucking stomach draws the liquid into the mouth.

Sex

Finally, let's look at *Phidippus'* sex problems. To do this, of course, you have to decide what sex spider you have. In some species, females look quite different from the males; in others, sex determination is mostly a matter of size. Females usually are larger than males, and their abdomens are as large or larger than their cephalothorax. Coloration may be different, usually with the males having more distinct markings. This latter trait relates to courtship, because males start courtship behavior and show off their "badges," much like peacocks display their tailfeathers. An excellent (and inexpensive) reference is *How to Know the Spiders*, 3rd edition, by B.J. Kaston, published by Wm. C. Brown Co. Armed with this book and the *Golden Guide to Spiders and Their Kin*, published by Golden Press in New York, you can identify most common spiders.

The *Phidippus* in my backyard is *P. audax*. It displays three bright white spots

gridded floor you can check out the following general scenario:

A jumper first orients itself toward the prey after detecting motion with its lateral eyes. Although its large anterior eyes focus to infinity, because of the spider's size, its maximum effective visual range is perhaps 30 to 40 cm. It will face its prey only when it notices moving objects. The prey will then face the stare of the two large anterior eyes. When its victim is 30 to 40 cm away the spider will begin stalking, and its final leap will cover 1 to 5 cm.

The prey is impaled with the pair of fangs. The pedipalps are initially used to feel the prey. However, the victim is held

Critical Communication

Figures 10-12A & 12B: A male jumping spider (*A*) waves his first pair of legs back and forth in species-specific patterns to insure he doesn't get eaten by the female. In the photo (*B*), a female Phidippus (left) observes the display of a male (right). (*A* redrawn from photo in Jackson, 1982)

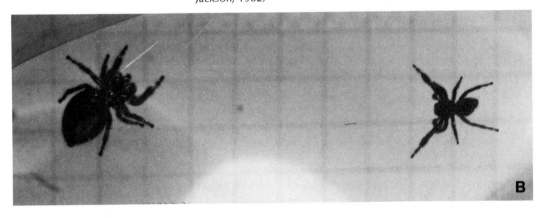

on the top of the abdomen with fainter markings arranged around them. Large females are about half an inch long, and males are somewhat smaller. The white hairs and scales on top of the palpi and the long first pair of legs bush out to frame the face like patches of whiskers.

If you have several spiders, the best way to tell sex is by watching their behavior. A pair of jumping spiders that meets in the open will first face each other. If a male decides he is in the presence of a female, he will rise up on tiptoes ("stilt"), sway a little from side to side, and raise his long first pair of legs. A receptive female will remain still and watch him closely. A female who has recently mated, or another male, or an animal of a different species, will respond by raising its legs. If this occurs, one or both animals will probably turn and run. But if the female remains quiet, the emboldened male may begin to "dance." The dance involves prancing and sidling in wide, circling arcs from one side of the female to the other. The arcs of this zig-zagging dance, with legs still extended, get smaller and smaller, bringing the male closer to his paramour. Nearly any motion on the part of the female may cause the male to bolt and run, but as long as she remains crouched and attentive, he will get closer.

This, undoubtedly, is the critical time for the male. Once he gets within the female's striking distance he is potential lunchmeat, for she is usually bigger than he is. No one knows for sure what the courtship accomplishes. Perhaps it serves to get the right species mated so that more young spiders will hatch alive and well; perhaps it serves to get the female aroused and ready, or perhaps it inhibits the female from wanting to eat the male. Maybe males that dance the right dance are displaying some sort of fitness that only a female can recognize. At any rate, females do seem to prefer a good dancer.

When the male gets within touching distance, about 2 to 3 cm, he will stop zig-zagging. He lowers his abdomen, shifts forelegs forward, and creeps ahead until he can touch and tap the female's head. If all is still going well at this point, he will mount the female and insert his sperm into her gonopore with his palp.

You can fool a jumper by showing him his reflection in a mirror. A male or female may react by spreading its first pair of legs in the threat/display posture. I've also seen a female jumper that spent long periods of time in front of the mirror posturing on tiptoes with her abdomen raised. Since only a few species have been studied in detail, there is still much to be learned about the behavior of these spiders.

More Sex

The male jumping spider, of course, doesn't always have his adventures on the road. When males find females in their nests a second type of courtship takes place that seems to depend more on mechanical and chemical cues than vision. A male engages in various tugging, probing, and vibrating movements on the surface of the nest. If the female inside is an unmated adult, they may then mate. If the female is a juvenile, the male will often spin a nest adjacent to her and wait until she has her final molt.

If you have a few jumping spiders in captivity for very long, you will see the type of nests they build. They are fairly thin and roughly cylindrical. The spider will either rest within or on it. The nests are not, as with orb weavers, designed for catching prey.

The Archnid Lesson

If spiders still haven't totally captured your sympathy, I'll hope you'll agree that their life style and behavior is more intricate than you thought. Arachnids have come a long way in the 400 million years since a scorpion-like creature ventured onto dry land. In number of species, the mites are most numerous, more than all of the rest of the arachnids put together. Over half of the remaining modern species are spiders and pseudoscorpions. If you take the time to look, you'll find many representatives tucked into the nooks and recesses of your home and backyard.

The next time you see a wandering wolf spider or startle a *Phidippus* into raising his forelegs, stop and listen closely. Perhaps you'll hear the murmurs of a warm Silurian sea.

REFERENCES

Bristowe, W.S. 1971. *The World of Spiders*. London: Collins Clear-Type Press. Most of this book is from text written in the 1950's, but it has interesting behavioral studies of the different spider groups.

Foelix, Rainer F. 1982. *Biology of Spiders*. Cambridge, Massachusetts: Harvard University Press. An excellent, detailed and expensive text on spiders in general.

Forster, Lyn. 1982. Visual Communication in Jumping Spiders (Salticidae). In *Spider Communication Mechanisms and Ecological Significance,* Peter N. Witt and Jerome S. Rovner, eds. Princeton, New Jersey: Princeton University Press. A detailed and technical text.

Greenberg, Joel, ed. 1984. Spider Silk Stretch and Strength. *Science News*, vol. 125, no. 25, p. 391.

Jackson, Robert R. 1982. The Behavior of Communicating in Jumping Spiders (Salticidae). In *Spider Communication Mechanisms and Ecological Significance,* Peter N. Witt and Jerome S. Rovner, eds. Princeton, New Jersey: Princeton University Press.

Kaestner, Alfred. 1970. *Invertebrate Zoology*, Vol. II. New York: Interscience Publishers.

Kaston, B.J. 1978. *How to Know the Spiders*, 3rd ed. Dubuque, Iowa: Wm. C. Brown Co. This book is the best now available for spider identification.

Levi, Herbert W. and Lorna R. 1968. *Spiders and Their Kin.* New York: Golden Press.

Peckham, G.W. and E.G. 1889. Additional Observations on Sexual Selections in Spiders of the Family Attidae. *Occasional Papers of the Wisconsin Natural History Society*, vol. 1, pp. 3–60. Mr. and Mrs. Peckham are the source of much original material on the behavior of jumping spiders. Their articles are in hard-to-find journals.

Preston-Mafham, Rod and Ken. 1984. *Spiders of the World.* New York: Facts on File Publications. A very visual text, with many four-color photographs.

11

A Tale of Two Kingdoms

"What sort of insects do you rejoice in, where you come from?" the Gnat inquired.

"I don't rejoice in insects at all," Alice explained.

Lewis Carroll's characters in *Through the Looking Glass* summarize many people's attitude toward insects. And yet, if we measure success by number and variety of forms, they are perhaps the most successful group of organisms on our planet. That success is undoubtedly one of our reasons for disliking many insects, because they compete with us for food, drink our blood, infest our bodies, and attempt to share our homes. On the other hand, insects have not singled us out. They have many long-term "arrangements" with members of the plant kingdom that work to our benefit. Insects are an important aspect of plant reproduction—particularly plants of the flowering variety that appeared some 120 million years ago as dinosaurs were taking their last bows.

It's difficult to tell the story of insects without including plants. It would be an artificial separation at any rate, because living things have always developed intimate interrelationships with each other. In the course of this coexistence, however, insects and plants have both pioneered certain innovations that allowed them to survive and flourish on dry land, including the fragment bounded by your

Figure 11-1: *Insects dominate the animal world by the number and variety of their forms and beetles are the largest order of insects. The two beetles shown here have an adversarial relationship. The black-and-red-eyed stink bug* (Perillus bioculatus) *on the bottom preys on the Colorado potato beetle* (Leptinotarsa decemlineata), *which is struggling to right itself (top).*

backyard.

Plants, for example, developed a waterproof coating and strong stems which allowed them to stand upright in dry air. Pores (stomata) in their trunks (and later their leaves) allowed gas exchange to occur. Spores, initially borne in terminally-placed capsules or strobili, were eventually carried on lateral shoots that allowed for tree-size growth.

Like amphibians, early plants were greatly dependent on water for reproduction. Spores germinated into small gametophyte plants that produced gametes. Male gametes needed water to reach their female counterparts before fertilization and the production of a new spore-bearing (sporophyte) plant could occur. Ultimately the sequence condensed itself until the gametophyte never left the spore-bearing plant and was surrounded by protective layers. Male sperm nestled in capsules called pollen grains and rode the winds or hitched a ride on arthropods that stopped by for a meal. When they found and fertilized the female gametophyte, the resulting sporophyte was then a seed, hidden safely in the branches of the previous sporophyte generation.

At first the seeds were relatively naked, housed in cones like those of modern conifers. As gametophytes became more cloistered, however, getting gametes together became chancier. Sugary rewards to attract potential arthropod pollinators became "cost effective" and so did showy petals to advertise the treats. Seeds became enclosed in fleshy tissue that also ripened and turned sweet and tempting.

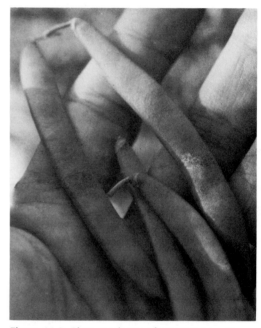

Figure 11-2: Plant seeds are often miniature time capsules that can transport a plant species across a long period of unfavorable conditions. These beans grew from Anasazi (ancient Pueblo Indian) bean seeds that had lain dormant in the dwellings of their cultivators for at least 700 years.

Animals ate the fruit, dispersed the seeds, and spread the successful flowering plants.

Insects went through a series of changes of their own that built upon a sound basic design. They inherited a segmented body from annelid ancestors. A modern-day intermediate between annelids and insects is exemplified by a creature called *Peripatus*. It survives in the tropics, eating some of its faster, more "advanced" brethren by tangling them up in the sticky threads it shoots out from openings on either side of its mouth. *Peripatus* bites its victim and injects enzymes which help predigest it. *Peripatus* is worm-like in general appearance, but its many legs are tipped with chitinous claws that presage the external skeleton of insects.

The external skeleton of insects provided anchorage for muscles needed to support animals on land. When coated with wax, the chitinous skeleton also provided protection from drying out, as well as a shield to help ward off predators. Breathing tubes called tracheae pierced the armor and provided a conduit for gas exchange. This last development did limit the size to which insects could grow, however, as there are limits to the efficiency of gas diffusion as animals get bigger, even when muscles are used to pump air in and out.

The first major insect innovation incorporated into this basic plan was flight. Primitive straight wings, like those of dragonflies, carried insects into an aerial domain unchallenged for 100 million years. Foldable wings, like those of the delicate, green lacewings followed. Finally, beetles gave up some of the grace of flying for the advantages of foldable wings covered by heavy, but protective, wing covers.

A simple process of molting, followed by growth and the production of a new and larger exoskeleton, was gradually superseded by molts followed by a period of dormancy as a pupa. During the dormancy, alternate genetic pathways were called upon, and the adult that emerged was radically different from its juvenile form. Metamorphosis provided the opportunity for immature and adult stages of an insect to utilize different parts of the environment, and to specialize in different aspects of the life cycle—such as growth and reproduction.

Ultimately, insects explored complex social organizations that, at least superficially, resemble those of vertebrates like ourselves. But whereas vertebrate societies are groups of genetically distinct individuals, most insect colonies are genetic duplicates of one individual—a "queen"—that have specialized in various functions.

Let's take a look at a plant-insect assemblage in your backyard that exemplifies these innovations.

The Milkweed Community

Walk outside and look for milkweeds. They should be easy to find. They're growing where you planted the petunias

Figures 11-3A, 3B, 3C, & 3D: Milkweeds are common in most of the U.S. This grouping (**A**), perhaps a clone of one original plant, competes with grasses along the fence. The thick, milky sap (**B**) is an apt inspiration for the plant's common name. **C** shows detail of the leaves and young flowers. **D** shows the flowers in bloom. Monarch butterflies are relatively famous for their use of the plant's glycosides to discourage predators, but many animals make a living on, near, or in the plant.

or the carrots. Milkweeds love to germinate in disturbed ground, and people are always disturbing some for them. Milk-weeds grow asexually when a seed finds virgin ground, so all the milkweeds in your garden may be clones of one plant.

Springtails

A

B

*Figures 11-4A & 4B: Two examples of springtails: (**A**) Hypogastrura nivicola and (**B**) Isotoma viridis. Springtails can be found world wide, from the Arctic to Antarctic regions. They feed on decaying material, algae, lichens, pollen and fungal spores. In general, those that live near the surface have well developed eyes, longer antennae and bigger "springs" (furculae). Subsoil forms may be white, eyeless and springless. Note the tube (collophore) on the bellyside of I. viridis. Collophores may be important in water absorption. (Redrawn from Bland,* How to Know the Insects, *1978)*

Milkweeds get their name from the milky sap that oozes out from their stems and leaves when you try to rip them from the ground. The family name for this group is *Asclepiadaceae*, taken from Asclepios, the Greek god of healing. Preparations of milkweed sap were considered to have various medicinal properties from time to time, and in fact, the sap has alkaloids which are bioactive.

Milkweeds are host to a variety of insects and other arthropods, and by looking at this one plant carefully, we can find examples of all the major insect and plant innovations that have occurred over the last 400 million years. It seems ironic that a plant you may consider as little more than a pest can provide such a vast perspective on natural history.

Springtails: A Casual Association

If you visit a milkweed plant on a sunny, relatively warm, winter day you may discover some of the most common but rarely seen insects: springtails.[1] Springtails are also some of the most primitive insects. Small and wingless, they live in leaf litter and soil. The stems of weeds poking through a layer of snow give them access to the surface. They often swarm in groups of half a million or more individuals. Springtails look like grains of pepper dusting the snow until you disturb them. Then they seem to blink in and out of sight and may appear on your hand if you hold it near them.

---

[1]Springtails are not considered true insects by some authorities. Their order, Collembola, is sometimes elevated to a separate class. Springtails have only six abdominal segments instead of the typical eleven; they possess the unusual furcula and a tube called the colophore attached to segment 1 that is unique to this group. Most important, their mouthparts are retracted into the head, unlike those of true insects.

Springtails achieve this magic with a forked organ called a furcula attached to the fourth abdominal segment. (Since springtails are less than six millimeters long, you need magnification to see any anatomical detail.) The furcula is folded forward under the abdomen when the animal is resting and is held in place by a structure on the third abdominal segment. The furcula is released and quickly snaps downward like a coiled spring, propelling the springtail as much as 100 mm into the air.

Springtails lay eggs in the soil during late winter and spring. The nymphs hatch and feed all summer. Nymphs and adults look the same except for size. They are not programmed for metamorphic changes. Large swarms may migrate over the snow for distances of 25 m which, considering their size, is quite a trek. Little is known regarding the whys and wherefores of this behavior.

Springtails can trace their ancestry back some 400 million years to the Devonian. In Rynie, Scotland, some of the first land plants and animals were preserved as fossils in a brittle, glassy material called chert. Simple vascular plants with forked stems tipped with sporangia (Psilophytes) and others resembling modern day club mosses lived in swampy areas and hosted tiny arthropods that nibbled at their stems and ate their spores. *Rhyniella*, the ancient springtail discovered there, had chewing mouthparts like its modern cousin and was either a scavenger or a plant eater.

To look at springtails under a microscope, pick up individuals with a brush dipped in alcohol or an aspirator. They can be preserved in isopropyl alcohol or 80 percent ethyl alcohol.

You can find another primitive, wingless insect without even leaving your house. Silverfish are slender insects with long antennae and three long tail-like appendages on their rear ends. They prefer dark and somewhat damp locations and so are partial to basement living. They lay eggs in available nooks and crannies. The young develop slowly through as many as a dozen molts. Like springtails, young and adults look the same. Silverfish move quickly, but you can catch some by putting out a box baited with crackers. Provide a cardboard ramp to the top of the box and smear the insides with petroleum jelly so they can't get out.

Looking for Dragons

Visit the milkweed plant in spring or summer and you may find another casual visitor, especially if you are near a pond or other body of water. "Deep in the sun-searched growths the dragon-fly/Hangs

Figure 11-5A: Dragonflies, one of the earliest orders of flying insects, shared their world with a different mix of plants than we have today. The larger plants were represented by ferns, seed ferns and gymnosperms. The enclosed seeds and flowers of angiosperms would not dominate the plant world until much later. The fern-like plant in the background is patterned after Medullosa noei. The plant in the right foreground with lobed leaves and a terminal seed is the extinct seed fern Neuropteris heterophylla. The fan-like leaves on the plant to the left belong to Ginkgo biloba, an ancient line of gymnosperms that still survives today. (Plants based on photos and reconstructions in Foster and Gifford, 1959. Dragonfly based on photograph by Dalton, 1975)

like a blue thread loosened from the sky" is how Dante Garbriel Rossetti described this creature. Dragonflies are not intimately tied to milkweeds, but may use them as a perch from which to strike out at a potential meal. Dragonflies are efficient predators—one of the first to exploit the air as a medium of transportation and living space.

Dragonfly wings are simple in basic design, but very effective. Dragonflies can travel at 30 miles per hour and lift up to fifteen times their own weight. The forewings are used to create a turbulent flow of air that

Figure 11-5B: Modern dragonflies have the same straight-wing design as their carboniferous forebears.

the hindwings can exploit for lift. Contrast this with the lift created by aircraft wings when air flows over a curved surface on the top of the wing and a relatively flat surface on the bottom. Since more air can slide beneath the wing than over it in a given amount of time, lift is created when the plane is moving fast enough. If engineers could successfully analyze the more complex use of turbulent air flow, they could create more maneuverable aircraft or at least reduce the vulnerability of existing planes to stalls.

Dragonflies and other insects pioneered their flying techniques 280 million years ago during the late Carboniferous—a period so named because the numerous terrestrial plant and animal communities of that time were preserved as the elemental carbon of today's coal fields. Under the tropical conditions of those times, dragonflies with two-foot wingspans cruised the skies, preying upon other winged pioneers beneath a canopy of scale-trunked trees over a hundred feet tall. It would have been a strange forest for us with no bird calls to punctuate the silence and no flowers to speckle the undergrowth. The whine of insects competed with the moaning wind and the croaks of amphibians courting in the night. Pendulous strobili swayed from high branches, releasing their spores in tenuous, drifting clouds.

Insects are unique among flying creatures in that their wings are not modified limbs, like those of birds and bats. They seem to have only the function of providing a means to fly. For many years scientists were hard pressed to explain how wings could have gradually evolved, because there would seem to be little value for the animal in partially formed wings. Recently, however, researchers have found that "protowings"—small outgrowths from the thorax—can have a significant effect in regulating body temperature (see Gould, 1985). The basal one third of *Colias* butterfly wings, for example, allows the animal a 55 percent greater increase in body temperature than an animal deprived completely of its wings. The temperature effect dwindles beyond a certain minimum length, however. Beyond that length the structure begins to demonstrate aerodynamic properties. Moreover, the domain of transition between thermal and aerial effects varies systematically with body size so that the larger the body, the sooner the transition in terms of relative wing length. Thus a primitive insect may have gained "flight readiness" by merely growing larger without major changes in body form.

Dragonflies are relatively hard to capture because their eyesight is excellent.

You'll notice that their huge eyes meet near the center of the head. Go after them with a net, approaching from the rear, and try to stay partially concealed. Their cousins, the damselflies, are somewhat easier to catch because they land on plants more often. Damselflies have more slender bodies, smaller eyes, and they hold their wings vertically when at rest instead of horizontally. Both are similar in their habits, however. Both members of the order (Odonata) also have aquatic, predatory nymphs that grow from 26 days to six years in the water (depending on species) before emerging as adults. The brightly colored hunting creatures that glitter on your weed stems have two to three weeks to mate and lay their eggs in the water before giving way to the next generation.

Lacewings

Most people encounter lacewings fluttering about a porch light on a warm summer evening. A common species has delicate, transparent wings, a slender, green body, a pair of hair-thin antennae, and golden eyes. However, don't be surprised if you find a few on your milkweed plants from time to time. Lacewings, both as adults and larvae, love to dine on aphids, and aphids in turn love to suck the sap of milkweeds.

When you find a lacewing at rest, notice that the wings are folded over its back. Folded wings represent an advance in flight versatility because the insect possessing them can move around in weedy growth with greater ease. Stoneflies and locusts are examples of some of the first groups of insects to demonstrate this characteristic, but lacewings are a particularly attractive example. Of course, aphids might not agree. Lacewing larvae hatch out of stalked egg masses as large-jawed predators, often referred to as aphid lions because of their taste for these tiny plant suckers. Lacewings are sometimes grown commercially to help gardeners keep aphids in check.

To see the life cycle in detail, place some adults in a coffee can or cardboard ice cream container. Replace the cover with a piece of cheesecloth held on with a rubber band. Place a moist cotton ball in the bottom to provide a source of water. Place the container out of direct light in a temperature regime of 70 to 80°F.

Adults will lay their eggs around the inside of the can and on the cheesecloth. The eggs will be white, oblong, and mounted on the end of a thin, rigid stalk. Using something like an emery board, you can scrape the sides of the can to deposit the eggs on a piece of paper. Transfer the eggs to a shoebox lined with tissue and

The Life of the Green Lacewing

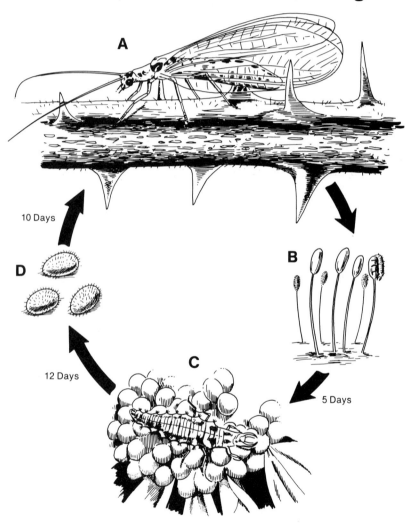

Figure 11-6: *The adult lacewing can be fed a protein and sugar diet. They live 20 to 40 days and lay 10 to 30 eggs per day. Lacewing larvae crawl out of their stalked eggs and feed on aphids. You can also feed them potato tuber moth eggs and larvae or grain moth eggs and honey or other soft-bodied insects. (See* Bugs: How to Raise Insects for Fun and Profit *by Daniel and Connie Mayer for information on raising other insects.)*

Figure 11-7A: *Tent caterpillars are common pests of fruit trees as well as aspen (the ones shown here) and poplars. These caterpillars are the larvae of lappet moths. Early morning or evening you will find most of the caterpillars in the "tent" of silk. During the day they forage on their host plant. These animals are insects of the order Lepidoptera, family Lasiocampidae and genus Malacosoma.*

Figure 11-7B: *The common mealworm (Tenebrio molitor), suggested as a food for many small creatures in this book, shows complete metamorphosis. Larval and adult stages are shown here.*

Some of the larvae will cannabalize others if they are crowded enough. Larvae will pupate in fuzzy, ovoid cocoons and hatch into adults in six to ten days.

Metamorphosis

The kind of development displayed by lacewings is called complete metamorphosis. The insect passes through four

evenly distribute the eggs on several layers of the tissue. Cover the shoebox with cheesecloth to keep the larvae inside.

Eggs hatch in five days and the larvae will feed for about twelve more. Give them aphids or other soft-bodied insects to eat.

stages: egg, larva, pupa, and adult. Ninety percent of all insects today undergo complete metamorphosis, although at the time it first appears in the Permian geological period, only 5 percent of insects used it. Obviously, complete metamorphosis must represent some real advantages for insects. Those advantages may be more apparent if we look at the

Incomplete Metamorphosis in the Chinch Bug

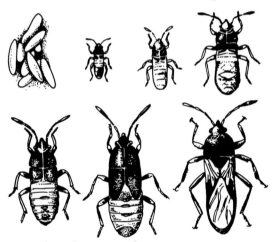

Figure 11-8: The chinch bug, like the large milkweed bug, belongs to the seed bug family (Lygaeidae). The chinch bug is more notorious, however, since its taste for corn and other grains more nearly coincides with our own. Note how the wing pads expand with each molt. In primitive insects, wing pad structures may have been important in temperature regulation before their aerodynamic properties were exploited. (Redrawn from Ross, 1956)

kind of development that preceded it.

The simpler form of development called, logically enough, incomplete metamorphosis, is displayed by other visitors to the milkweed plant. Dragonflies and damselflies display it, although during their immature stages they are aquatic and possess gills and other specializations that make them somewhat atypical. The milkweed bug, *Oncopeltus fasciatus*, however, is a good example. The milkweed bug likes milkweed seeds and dissolves them to an edible state with his saliva. The adult bugs are fairly large, 10 to 18 mm, dark brown to black, with two orange bands across the wings and orange markings on the head. You can distinguish males from females quite easily if you look at their underside. Females have black spots on the two posterior abdominal segments, whereas the males have a black spot on only the last segment.

When young milkweed bugs hatch from their eggs, they look like miniature adults without wings. Where the wings would be attached, on the second and third thoracic segments, are short wing pads. The young bugs eat seeds and have to compete with older members of their own species as well as any other seed-eating competitors. Insects can grow only so much within the confines of their chitinous skele-

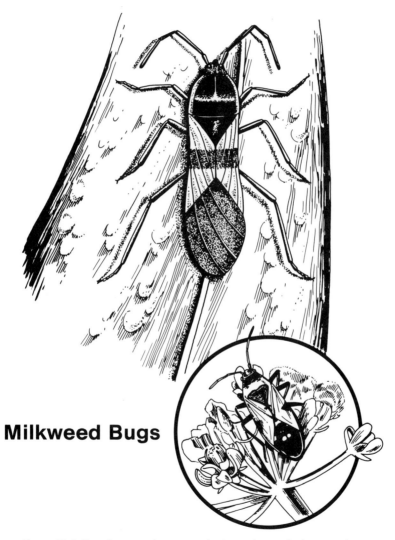

Milkweed Bugs

Figure 11-9: *True bugs, as the entomologist understands the term, belong to the order Hemiptera. Two bugs like to feed on milkweed. The larger,* Oncopeltus fasciatus, *dines on seeds and is shown on a seed pod. The smaller,* Lygaeus kalmii, *is also a seed bug and is shown on a flower head. Both are drawn to the same scale.*

tons and then they must shed them to allow more growth. Insects go through several of these molts, the exact number depending on the species. The milkweed bug gets bigger and the wingpads longer at each successive molt until, after the last molt, the winged adult emerges. Thus, in incomplete metamorphosis, there are three stages—egg, larva, and adult—that develop one from the other in a rather straightforward progression. Cockroaches, grasshoppers, stoneflies, and leaf hoppers are other backyard insects that have this kind of development.

In the lacewing, of course, the voracious aphid lion larvae look nothing like their delicate adult forms. The differences are acquired during the inactive pupal stage that was absent in milkweed bug development. During this stage whole sets of genes are turned off and alternate ones turned on. Fat supplies are mobilized, muscles are broken down and rearranged, and suppressed zones of cells are activated and allowed to multiply. The resulting adult could easily be mistaken for a whole different creature. What could have selected for such a drastic reorganization?

One key factor may well have been the competition for limited resources. All stages of milkweed bugs feed on milkweed seeds. If larvae and adult were radically different, however, different selection pressures could act upon them.

There would, in fact, be great advantages accrued by specializing in ways that would take young and adult out of direct competition with each other. The monarch butterfly is a milkweed guest that serves as a classic example.

The Kings With Bad Taste

Monarchs lay their eggs singly on milkweed plants. In four days larvae hatch out and begin munching milkweed leaves. Many animals avoid milkweeds because of the potent alkaloids they produce. Although this has, on the whole, been a good defense measure on the part of milkweeds, monarchs have adapted to the poisons and even turned them to their own benefit. Most birds and other animals avoid monarchs because, as revealed in classic studies with young birds, the latter vomit after their first meal of monarchs and remember the experience. More recent studies show the milkweeds have become less poisonous due to agriculture practices, and birds have learned that the poison is only in the outer body wall area. They scoop out the insides of adults and larvae and leave the poison behind.

Monarch butterflies molt four times and eat voraciously for ten days between molts. After the last molt they attach to the

The Monarch Life Cycle

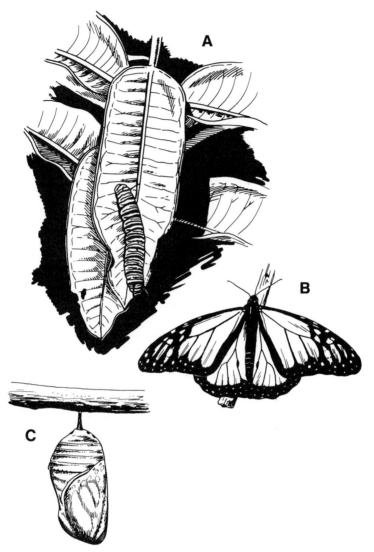

Figures 11-10A, 10B & 10C: *The monarch is a classic example of both complete metamorphosis and warning coloration. Birds and other animals learn to avoid monarchs because of the alkaloids they incorporate into their bodies from milkweed plants. Even the pupae (**C**) are poisonous. The larvae (**A**) is shown feeding on a milkweed frond. The adult (**B**) can be confused with the viceroy which has a band of black crossing the hind wings perpendicular to the veins.*

underside of a leaf or other firm support and pupate. Adults emerge in 12 days. Many of us have witnessed this aspect of the butterfly life cycle. The adult that hatches advertises its dangerous taste with bold orange and black colors. More importantly, it is now equipped to eat in an entirely different fashion, so the adult and larval stages will not be competing for the same food source. The monarch has traded in its jaws for a slender tube that can snake down within a flower and draw out nectar. You can observe this by offering a captive adult a little sugar water.

Reduced competition for food is not the complete answer to the success of complete metamorphosis. The value of specialization is another factor. Caterpillars are eating machines. They eat continually to build up the raw materials that will later be converted to black and orange beauty. All their energies are concentrated on this activity, for their success is based on feeding efficiency rather than a compromise that balances the requirements of feeding and reproduction. Adult monarchs, on the other hand, specialize in reproduction. They eat high-energy sugar snacks to get them through to egg-laying time or fall migration.

You may see adult males perched on milkweeds or other plants. When a female approaches, the male will fly up to inspect her abdomen. If the match is right, there will be a short chase, followed by a descent into grass or weeds. The male will attach his abdomen to the female's and may fly off. The female will retract her legs and fold her wings. Sperm will be transferred during the next two to fourteen hours.

Monarchs are easily confused with other butterflies called viceroys. Both are orange with black-veined and black-rimmed wings. Viceroys, however, have an extra black stripe that runs across the other veins on the bottom pair of wings. Viceroys are rarer than monarchs and have successfully copied the latter's coloration to discourage predators. Viceroys have no taste for milkweeds and are found on

Figure 11-11: *The admiral butterfly (*Limenitis arthemis*) is a close relative of the viceroy (*Limenitis archippus*). The latter species mimics the monarch butterfly and thus benefits from the monarch's "bad taste." The admiral has no such protection.*

June Beetle in Flight

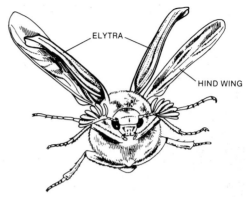

aspens, poplars, and willows. They overwinter as partly developed larvae on rolled up leaf fragments. Monarchs migrate to Mexico or southern California in the fall and roost by the millions in mountain pine trees.

Masters of the Earth

Emily Dickinson claimed that "success is counted sweetest by those who ne'er succeed." If true, beetles must be quite blasé about their accomplishments. Approximately one out of every four animals on the planet is a beetle. Many people have no great fondness for beetles, either because they take exception to their looks or resent their attempts to share our cereal grains. The Egyptians went so far as to deify the scarab beetle, however, and portrayed the god, Khepri, as a scarab rolling the sun across the sky. Every day the sun would "die" in the West and be "reborn" in the East, just as young scarab beetles rose from balls of dung which the adults had carefully buried.

Beetles represent the most recent refinement in insect flight: the foldable, covered wing. They aren't the most graceful flyers, but they fly well enough to take care of reproduction and dispersal. When they aren't flying, their wings are folded

Figure 11-12: Although they look like an airborne truck, insects like the June beetle are very successful. Flight is for dispersal and reproduction. At other times the flying wings are safely tucked beneath the hard forewings (elytra) so that beetle can exploit ground and subterranean habitats. (Drawing based on photo by Dalton, 1975)

neatly away in protective armor. Actually, their forewings are the armor. They are usually hard and rigid, often with parallel ridges, not unlike the design of some modern day fiberglass materials. Beneath these wings, called elytra, the hindwings are folded and protected like a parachute in a backpack. When closed, the elytra of most species fit together snugly with interlocking hooks.

Beetles have other things going for them, too. They are fast and agile ground dwellers with general-purpose, chewing mouthparts. They are good at water conservation; have developed a variety of protective devices involving chemistry, camouflage and behavior; and have life histories that tend to keep their larval and

Figures 11-13A & 13B: Convolvulus arvensis (*A*), commonly called bindweed, is a pest plant in Colorado that chokes out most competitors. If, as a gardener, you can be dispassionate, the flowers are rather attractive and resemble their domestic cousins, the morning glories. The larvae of tortoise shell beetles (family Chrysomelidae) feed on bindweed (*B*). With a forked tail, these larvae hold old skin and excrement over their bodies like a parasol as they eat. Presumably this is a form of concealment. As adults, these small, nearly round beetles have transparent or translucent elytra (wing covers) that appear metallic gold when they are alive.

pupal stages under cover. The ever-present milkweed can offer yet another example.

The milkweed beetle, *Tetraopes tetraopthalmus*, is one of the longhorned beetles, characterized by long antennae with archlike horns on the head. The milkweed beetle is 12 to 15 mm long, red with black spots and black antennae. Like monarchs, it can and does feed on milkweed leaves. The eyes are divided, for no obvious reason, so that there are four eyes instead of the normal two. *Tetraopes* larvae bore into the milkweed stem and enjoy a private dining room. Other mem-

bers of this beetle family (Cerambycidae) are called round-headed wood borers. The adults lay eggs on weak or fallen trees. Larvae are cylindrical in shape, whitish, and have a rounded head and thorax area.

Beetles arose 230 million years ago and very quickly accounted for 40 percent of all insects. In addition to all the positive features already mentioned, part of this success may have resulted from just being in the right place at the right time. They successfully exploited ground and sub-surface habitats that offered many niches for specialization—no small number of which are in your backyard.

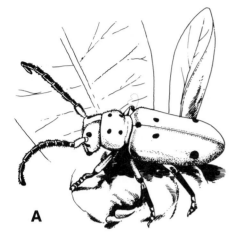

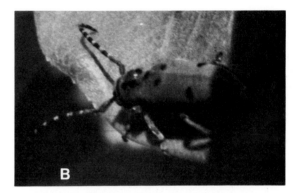

Figures 11-14A & 14B: *Milkweed longhorn (*Tetraopes *sp.) larvae bore into milkweed stems and eat in relatively well protected peace. The adult is red with black spots and antennae.*

The Milkweed's Payoff

You've now seen several insects that live off milkweed plants, but so far you've seen no evidence that milkweeds can do anything except defend themselves with noxious chemicals. In fact, milkweeds have developed techniques that exploit some insect visitors to aid their own reproduction. Milkweeds are angiosperms, or flowering plants. The earliest fossil evidence of flowers extends back 120 million years to the Cretaceous period. Even at that time, the large size, radial symmetry, and prominent display of the flower on the plants all imply that insects were important in pollination. None of these early flowers, however, were as complex in design as the flowers of the

milkweed, which depend on heavy-bodied fliers to accommodate their needs.

A typical flowering plant carries its eggs (ovules) in the broad base of a flask-shaped structure called the pistil. The neck of the flask is called the style; the lip is called the stigma. Pollen grains, housing male nuclei, are borne on stamens, consisting of a slender filament topped with a head not unlike a hotdog bun in shape. When pollen grains land on the sticky surface of the stigma, a pollen tube grows down through the tissue of the pistil toward the ovaries. The male nuclei migrate down the tube to fuse with the eggs, from which union seeds develop. In milkweeds, however, the style of the pistil is short and the stigma is enlarged into a

Figure 11-15: Some species of longhorn beetles specialize in eating the pollen of flowers. Male and female of one species are taking a feeding break for copulation on this sunflower (top) while another species continues feeding (bottom). These longhorns have long snouts and broad shouldered, tapered bodies that allow them to get at pollen easily. After this flower sat in a jar for an hour or so, many smaller species of beetle crawled out, testifying to the multitude of relationships between insects and flowering plants.

deep slit. The pollen is not shed in single grains or small clusters but in large structures called pollinia in which the pollen grains are held together with wax.

The milkweed flower has five parts, each with a stigma and a pair of pollinia housed in sacs that are slung over the stigma like saddle bags. The connecting "straps" are thin bands called translator arms, held together by a "clip" called the corpusculum. The whole saddlebag unit is called a pollinarium.

The problem a milkweed plant has is that it must breed with another individual to get fertile seeds. As mentioned earlier, most of the milkweed plants in a given clearing may be part of one root-propagated clone. Pollinaria may have to be carried miles to the next clone. The problem becomes one of getting insects to volunteer their services as sex cell exporters. The milkweed solves the problem by bribing them. The bribe is a general one, consisting of a copious supply of nectar. Monarchs use the gift, but carry virtually no pollinia. Bumblebees, however, which are ten or twenty times more apt to visit the plant, are often found with strings of pollinaria on their feet. Their legs hook in the slit of the clip (corpusculum) and lift the saddlebag pollinarium off the stigma. As they fly, the translator arms rotate so that a pollinarium will fit neatly into the stigma of a subsequent flower. The pollinarium usually breaks off at the clip, and quite often other pollinaria hitch a ride. The bee may end up with a cumbersome load. Bees also lose legs in the process, and lighter insects such as flies may become trapped and die.

Bees aren't the only pollinators, however. Milkweeds hedge their bets. Milkweeds produce even more nectar at night

A Basic Floral Design

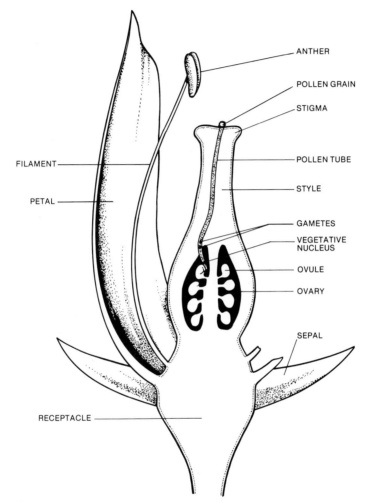

ANTHER

POLLEN GRAIN

STIGMA

POLLEN TUBE

STYLE

GAMETES

VEGETATIVE NUCLEUS

OVULE

OVARY

SEPAL

FILAMENT

PETAL

RECEPTACLE

Figure 11-16: *This cross-sectional view portrays a "typical" dicotyle-donous flower. When the pollen grain lands on the stigma a pollen tube grows downward under the direction of a vegetative nucleus. When it reaches an ovule, one of its gametes will fuse with the egg cell in the ovule. (Based on drawings in Meeuse and Morris, 1984)*

than during the day, and moths are at-tracted to them. White flowers are typical of night-pollinated species because of their high visibility. You may have already noticed that milkweeds are quite fragrant at night. Brown moths like *Catocala ilia*

The Milkweed Flower and Pollination

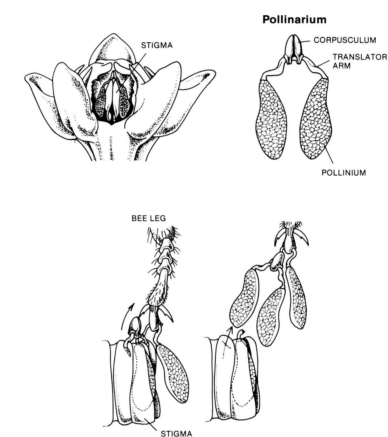

Pollinarium

STIGMA

CORPUSCULUM

TRANSLATOR ARM

POLLINIUM

BEE LEG

STIGMA

Figure 11-17: *The milkweed flower has an enlarged stigma and a complicated delivery system for its pollen. See text for explanation. (Redrawn from Morse, 1985)*

seem to account for about 5 to 25 percent of pollinations. They may be more important in cross-pollinations than bees because they are wide-ranging and not tied to a central home base.

Social Climbers

Bumblebees represent the last great insect innovation: social behavior. They belong to an order, the Hymenoptera,

Notes on Raising Insects

Note what particular insects are eating and where they are living when you find them. If you can create "a little bit of home" inside your house, you may be able to raise them successfully. Here are some points to keep in mind.

1.) You can make water available by making a cotton wick and putting it in a small vial of water.

2.) A small wire cage can be placed over outdoor plants to make a kind of vivarium.

3.) Some insect larvae need soil or twigs to complete their life cycle.

4.) Many insects are reared on substitute foods or artificial diets. The following are examples of foods used for some commonly reared insects:

Crickets: Dry powdered milk mixed with crushed, dry dog food; bits of apple, banana, or lettuce suffice for very young crickets.

Grasshoppers: Lettuce; powdered dry milk (two parts), dry alfalfa meal (two parts), and dry brewer's yeast (one part).

Cockroaches: Crushed, dry dog or rat food mixed with dry powdered milk.

Mantids: Fruit flies and other flies, aphids (leave them on the plant), crickets, and grasshoppers. Mealworms and uncooked hamburger or other meat will be eaten if held with forceps and touched to the mantid's mouth.

Large milkweed bug: Milkweed seeds (collect in the fall and store seeds for later use). A sunflower-feeding strain is also available from supply companies.

Mealworm (beetle): Bran with an occasional apple core added.

Butterflies and moths: Sugar water.

Mosquito larvae: Finely crushed, dry dog food sprinkled very lightly on the water which contains the larvae.

(Adapted from *How to Know the Insects*, 3rd ed., by Roger G. Bland and H.E. Jaques.)

that also includes wasps and ants. The story of the honeybee has been well told, but the bumblebee life style may be closer to the primitive beginnings of social behavior during the Permian.

Bumblebees overwinter as fertile adult

Figure 11-18: Flies are often important pollinators of mountain wild flowers.

Figure 11-19: Ants will "milk" aphids by lapping up the beads of sugary liquid the aphids expel from their rear ends. The ants, in turn, will defend their small charges from predators.

females. They come out of hibernation in the spring, after flowers have bloomed. If you come across a queen bee in the spring, you may be led to the nest site by carefully watching her weaving inspection of the leaf litter. These spring queen bees are the largest you will see until the fall. They seem to have a louder-than-average buzz as they cruise at low altitudes looking for a nest site. If they find an old, abandoned rodent burrow or other suitable site, they collect a supply of pollen, lay their eggs upon it, and cover both with a layer of wax. They sit on the eggs until they hatch in four or five days. The larvae feed on the pollen for a week before they pupate and emerge as sterile females. These females will care for all other young produced by the queen. In the fall, fertile males and females are produced. After mating, the new queens overwinter. The remainder of the colony dies during the winter.

Competition for nest sites is often acute. Perhaps 10 percent of nest sites are "take-overs" of one queen by another. The losing queen is either killed or becomes submissive to the victor.

Figure 11-20: *Winged aphids are infesting this columbine. Winged females fly from plant to plant and give birth to live young which develop into wingless adults. These adults, in turn, duplicate themselves for a dozen generations or more without males—a process called parthenogenesis. Eventually, a female produces males and females who mate and produce eggs that will hatch into nymphs in the spring. When you see aphids, look for predators like lady bird beetles, lacewings, or syrphid flies. Ants are also likely to be around milking aphids for their "honeydew." Ninety percent of aphids are restricted to one kind of host plant.*

The above description is typical of the genus *Bombus.* Another genus of bumblebees, *Psithyrus*, is a pirate that parasitizes *Bombus* nests. *Psithyrus* has no pollen-carrying hooks on its legs, and all young produced are males and fertile females. It emerges later in the spring when *Bombus* nests are in place. Invading queens kill

Bombus queens or destroy all eggs they produce. Their own eggs are laid in the hive and cared for by the sterile *Bombus* females.

Tales Untold

The milkweed story involves many other insects and arthropods, such as the crab spider that feeds on unwary visitors, daddy longlegs (harvestmen) that clean up the spoils, and several other species of moths. Ants may visit milkweeds to "milk" the aphids of a sweet-smelling liquid that is secreted from the anus. They also fight off lady beetles and other aphid enemies. The interrelationships are complex and fascinating, reflecting the long-shared mutual history of plants and insects that has allowed them to "pay off" each other to accommodate their needs. My advice is to pick a plant, any plant, and settle back to watch the show. If your neighbor should complain about the weeds, tell him you're studying the problem.

REFERENCES

Arnett, Ross H., Jr. and Richard L. Jacques, Jr. 1981. *Simon and Schuster's Guide to Insects.* New York: Simon and Schuster.

Arnett, Ross H., Jr. and Richard L. Jacques Jr. 1985. *Insect Life.* New York: Prentice-Hall Books.

Bland, Roger G. and H.E. Jacques. 1978. *How to Know the Insects,* 3rd ed. Dubuque, Iowa: Wm. C. Brown Co.

Clarke, Kenneth U. 1973. *The Biology of the Arthropoda.* New York, London: Edward Arnold Ltd.

Dalton, Stephen. 1975. *Borne on the Wind.* New York: Reader's Digest Press, E.P. Dutton and Co.

Dilcher, David and Peter R. Crane. 1984. In Pursuit of the First Flower. *Natural History,* vol. 93, no. 3 (March).

Evans, Glyn. 1975. *The Life of Beetles.* New York: Hafner Press.

Evans, Howard Ensign. 1968. *Life on a Little-Known Planet.* New York: Dell Publishing Co.

Foster, Adriance S. and Ernest M. Gifford, Jr. 1959. *Comparative Morphology of Vascular Plants.* San Francisco: W.H. Freeman and Company.

Gould, Stephen Jay. 1985. Not Necessarily a Wing. *Natural History,* vol. 94, no. 10 (October).

Grant, Susan. 1984. *Beauty and the Beast: The Coevolution of Plants and Animals.* New York: Charles Scribner's Sons.

Headstrom, Richard. 1982. *Adventures With Insects.* New York: Dover Publications.

Kingsolver, Joel G. and M.A.R. Koehl. 1985. Aerodynamics, Thermoregulation, and the Evolution of Insect Wings: Differential Scaling and Evolutionary Change. *Evolution,* vol. 39.

Lewin, Roger. 1982. *Thread of Life: The Smithsonian Looks at Evolution.* Washington,

D.C.: Smithsonian Books.

Mayer, Daniel F. and Connie. 1983. *Bugs: How to Raise Insects for Fun and Profit.* South Bend, Indiana: And Books.

Meeuse, Bastiaan and Sean Morris. 1984. *The Sex Life of Flowers.* New York: Facts on File.

Morse, Douglass H. 1985. Milkweeds and Their Visitors. *Scientific American,* vol. 253, no. 1 (July).

Peterson, Ivars. 1985. On the Wings of a Dragonfly. *Science News,* vol. 128, no. 6 (August 10).

Read, Morley. 1985. Stalking the Collared Peripatus. *Natural History,* vol. 94, no. 9 (September).

Reitter, Ewald. 1961. *Beetles.* New York: G.P. Putnam's Sons.

Ross, Herbert H. 1956. *A Textbook of Entomology,* 2nd ed.. New York: John Wiley and Sons, Inc.

Shear, Wm. A. et al. 1984. Early Land Animals in North America: Evidence From Devonian Age Arthropods From Gilboa, New York. *Science,* vol. 224, no. 4648 (May 4).

Stokes, Donald W. 1983. *A Guide to Observing Insect Lives.* Boston: Little, Brown and Co.

12

A Tiger in Your Window Well?

All the way home I knew I was in trouble. The rain was coming down so hard and fast that I knew my window wells were in danger of filling up. The water would then cheerfully squirt through the windows and into my basement. I was right, of course, as a person tends to be in the case of disasters, so while my wife denuded the house of towels and piled them beneath the window on the inside, I slogged around in typical Colorado clay soil, baled as much as I could out of the well, and rearranged the local topography with my shovel.

Downstairs I rigged a siphon arrangement that carried excess water into the drain. As I stood looking at my achievement, tired, but just a little self-satisfied, as only an amechanical person can be who has just triumphed over a "thing," a pair of eyes looked back at me. Dogpaddling in the window well was a large, colorful salamander, its broad snout bumping against the glass.

I discovered that the Western tiger salamander, *Ambystoma tigrinum*, is common in this part of the country. Its banded brown and yellow coloration and six-inch length make it quite distinctive. Many people don't see it, however, because it travels by night or during rainy periods in search of a mate, a place to lay eggs, or both. Water is as critical for salamanders as it is for more commonly seen amphibians like toads and frogs. These animals represent an early vertebrate experiment in land-living, and like the pill bugs, spiders, and insects before them, they had some tough problems to solve.

Vertebrate Solutions to Living on Land

You'll recall some of the formidable problems to be solved in taking up residence on dry land: air isn't nearly as buoyant as water, so organisms need to be preadapted with a rigid framework of some kind; the problem of acquiring oxygen must be solved without drying out in the process; organisms have to get from place to place in a totally different medium; they have to accomodate a much more severe range of temperatures; and they have to be able to successfully reproduce.

Like the invertebrate invaders of land, amphibians were preadapted in several ways that made the transition to land easier. The earliest amphibians, of which there are no living relatives, were the labyrinthodonts. The skulls of these amphibians are very similar to certain kinds of fish that lived in the Devonian, some 375 million years ago. These were sturdy fish,

Figure 12-2: The Western toad, Bufo boreas, is a common garden toad in the West. It has a pale stripe down the middle of its back, but no cranial crests like some other species. They are gray or greenish with warts set in dark blotches, often tinged with rust. They'll burrow into loose soil for shelter or use the burrows of small mammals. They'll eat about any insect of the right size.

Figure 12-1: This tiger salamander is coming ashore near primitive plants like the tall horsetails (Equisetum) and liverworts (Marchantia, in right foreground) that existed during the carboniferous when amphibians were the penultimate land vertebrates.

somewhat like modern lungfish, and like them they may have obtained some of their oxygen from the air. Lungfish are found in Africa, South America, and Australia. They are slender small-scale fish that grow to about 50 inches in length. When the water they live in dries up, they dig a hole in the mud and lie dormant until water returns. Lungfish have rather slender pectoral fins, but the amphibian ancestor

may have had well-developed pectorals like the modern-day mudskippers. These small fish, living in shallow coastal waters, spend considerable time jumping about on mud flats, eating insects, and breathing with the help of air trapped in their gills.

Many of the earliest amphibians were more formidable than those left today. Some, like Seymouria, were carnivorous and had rather impressive looking teeth. Modern amphibians fall into three groups whose origins go back several hundred million years into the Carboniferous. The Order Urodela is made up of newts and

Eryops
An Early Permian Amphibian

Figure 12-3: *Eryops was an early amphibian that is found occasionally in Permian rocks of the United States. Its short legs had to support a broad body and large head.* Eryops *probably didn't stray far from water. (Drawing based on photograph of a model in* Prehistoric Animals *by Ellis Owen)*

salamanders. Frogs and toads belong to the Order Anura. The tropics shelter some burrowing, legless amphibians belonging to the Order Apoda. The earliest amphibians may have had a scaled skin. While the Apoda have small, virtually microscopic scales, the rest of modern amphibians have scaleless skins and no claws on their toes.

Tiger salamanders, like their lungfish forebears, get along very well by lying dormant in the mud. In fact, these salamanders and their kin are called mole salamanders. The first tiger that stumbled into my window well I released into a small lake so that he could fulfill his salamandrian existence. Several years later, another tiger turned up during one of my dog's February excavation projects. My children alerted me to the creature that had been uncovered. The stiff and dusty lizard-like form looked more like a rubber toy than a real animal, but it warmed up quickly in my hand and waited out spring in a small aquarium.

Salamanders reveal the basic vertebrate

Amphibians
Ancient and Modern

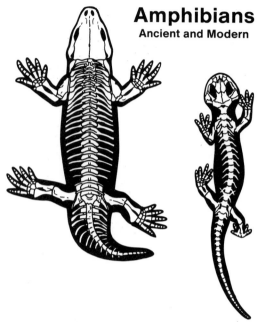

Figure 12-4: Amphibians demonstrate the basic vertebrate body plan of four legs and a tail. Their legs project from the sides of the body, however, rather than supporting the trunk from underneath as in reptiles and most higher vertebrates. The skeleton of Seymouria, a Permian amphibian, is shown on the left. It had features in the skull, limb girdles and digits like the reptiles that would follow. The skeleton on the right is a tiger salamander. (Seymouria based on a drawing in The Life of Vertebrates *by J. Z. Young and salamander based on a drawing in* Five Kingdoms *by Lynn Margulis and Karlene Schwartz)*

legs to move forward in a swimming fashion. The sinuous motions are not unlike those of its fish ancestor. A salamander out for a more leisurely walk will lift its legs with more deliberation. The gait, however, is very basic: the forelimb on one side will go forward as the hindlimb on the same side goes backward. If you pay attention to the swing of your arms as you walk, you will see the remnants of this basic vertebrate pattern.

One might think that lifting the body off the ground would be important in improving locomotion, and it probably was—secondarily. Primarily, amphibians needed to take some pressure off the lungs for efficient breathing. Amphibians have no muscular diaphragm as we do for expanding and contracting the chest cavity. Instead they use their mouth as a bellows. If you watch a frog or a salamander at rest, you will see them lower the bottom of their mouth periodically. When they do this, air is drawn in through the nostrils. They then close their nostrils from the inside with muscular valves. When the mouth cavity is returned to its minimum size, air is forced into the lungs. By lowering the mouth cavity and keeping the nostrils closed, air can be drawn from the lungs. Sometimes the same air is forced into the lungs several times to extract as much oxygen as possible. The same basic force-pump system is used by

body plan much better than the frog that you may have encountered in a biology class. Salamanders have long bodies with a tail and four limbs. They look like creatures caught in the middle of a push-up, because their limbs are out to the side rather than under the body. When you scare a salamander and it wants to move quickly, it will use its belly as much as its

Figure 12-5: *The basic "push-up posture" of simple vertebrates is illustrated well by the salamander.*

fish to force water over the gills and by larval salamanders.

Lungs, however, are sometimes not essential to amphibian survival. The amphibian skin is the most important breathing apparatus and is richly endowed with blood vessels. It must stay moist to work properly. Some salamanders in fast-moving streams, where oxygen is plentiful, lack lungs altogether. Extra lung capacity is especially useful at breeding time, however, and—except for eating—that's a salamander's main business.

The Migrations of Spring

The salamanders that fell into my window wells undoubtedly considered them private pools suitable for romance.

Though they were mistaken in the two cases I encountered, it's apparently not uncommon in the Midwest and West to find tiger salamanders in an assortment of temporary pools. In fact, in semi-arid and desert habitats it may be the only time you see adults. Although it's not yet known exactly what drives salamander migrations, in the case of at least some salamanders it is a strong and goal-oriented drive. One study of homing behavior in a species of river newt in California found that the newts could find their way back to a specific stretch of their home rivers after traveling several miles overland—even if it took them several seasons. They returned in very high percentages, with mortality apparently claiming a larger number as they were transported farther away. Perhaps, as in pigeons, bumble bees, dolphins, and even bacteria, there are deposits of magnetite tied into their nervous

Salamander Tracks

Figure 12-6: *These tracks were made by a tiger salamander moving on a smoked drum (the original tracks were white on a black background). On the left there was fast movement with the body on the ground. On the right, the salamander was walking slowly, raised up on its legs. (Based on the work of Evans, 1946, as illustrated in* The Life of Vertebrates *by J. Z. Young)*

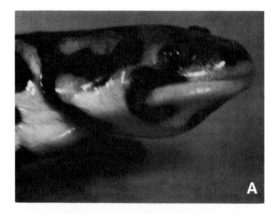

Figures 12-7A & 7B: *With nostrils open, the tiger salamander can draw air into its mouth by using jaw muscles to enlarge the mouth cavity (**A**). The salamander then closes its nostrils and decreases the size of its mouth cavity (**B**) which forces air into the lungs.*

systems so that they can somehow orient to the Earth's magnetic field, or perhaps this remarkable behavior can be explained entirely by some aspect of salamanders' acute sense of smell.

Male salamanders have special hedonic glands at the base of their tails and on the underside of their heads that produce odors females respond to. Females produce characteristic scents in their general skin secretions. Male tiger salamanders congregate at breeding sites before females and can occasionally be found in large numbers. Males are hard to tell from females. Their tails are longer, relative to the rest of their body, than females, but there is some overlap. In a Kansas population the male tails were 42 to 50.7 percent of their total length, whereas the female tails were 35 to 46.5 percent of their body length. After the females arrive, the males can begin their "dances of love," or liebesspiels. These dances involve a sinuous intertwining and emitting of perfumes by the females. When this has gone on for some time, the male finally deposits a jelly-like mass called a spermatophore, which contains his sperm. If the female has been sufficiently stimulated, she will pick up the sperm-filled cap of this spermatophore with the lips of her cloaca. Sperm can remain viable in special chambers within the cloaca for days or even months.

Eventually the sperm fertilize the female's eggs, which she lays singly or in small clusters along the stems of water weeds. For this aspect of their lives most amphibians are totally dependent on water or very moist conditions. There are some terrestrial salamanders, however, that lay their eggs on land. In these species, usually one or both parents stick around to help keep the eggs moist and protected. A few species give birth to live young that are small, functioning copies of the adults.

Most amphibian eggs, however, including those of the tiger salamander, fend for themselves, protected only by the jelly-like coating given them by their mother when they pass through her oviducts. Some freeze, some are eaten, and some are attacked by molds, but a good percentage hatch to release a larval creature that could be mistaken for an entirely different kind of animal. The process of development that ensues, culminating in the amazing transformation into a land-based organism, is one of the most fascinating aspects of studying these animals.

Forever Young—Sometimes

Most people are familiar with the classic transformation of tadpole into frog. Frogs' eggs hatch into brown, black, or mottled tadpoles that munch on vegetation, grow, and eventually trade in their tails and other fish-like characteristics for legs and a taste for insect cuisine. The salamander experience can be very similar, but there is also a series of variations that allows you to glimpse something of the origins of this lifestyle in distant vertebrate relatives.

Alternation between immature and adult versions of animals is not an uncommon strategy in the animal kingdom. Many insects display a dramatic change between larva and adult. The change is triggered by environmental cues and coordinated by hormone systems that selectively turn on and off different sets of genes. This allows young and adult versions of the same organism to exploit different resources and allows seasonal specialization in food gathering and reproduction. Alternation of generations in simple plants accomplishes similar things. This also means, however, that the different phases of such an organism's life cycle are under different selection pressures. Sometimes the immature form might be better suited to conditions at hand, and sometimes the adult version is favored. It's not too surprising, then, that conflicting signals from the environment may at times result in an immature form that doesn't make it all the way to "adulthood."

Retention of juvenile traits into adulthood is termed neoteny, and many salamanders display the trait to one degree or another. Variations run the entire gamut, from salamanders that always transform from a larval to adult form, to those that never do. Tiger salamanders fall into that interesting middle group where the larvae may or may not transform into land-based creatures. An untransformed tiger salamander is commonly called an axolotl, a name given to the species found near Mexico City that normally stays aquatic and "mostly juvenile" all its life. Axolotls retain their gills and fish-like shape, but do have small legs and reach sexual maturity while in their "larval" form.

If you give a dose of thyroxine to an axolotl, though, it will undergo the entire metamorphosis into a very tiger salamander-style adult. The transformation process requires substantial reorganization of the body plan, including loss of the gills, closure of the gill slits, appearance of a tongue pad, and reorganization of the gill skeleton and the attendant muscles to produce a tongue, enlargement of the mouth and eyes, development of eyelids, and major changes in the skull and skin. Thyroxine would appear to be quite a wonder drug, and as one of the main effector hormones in vertebrates, it probably qualifies. The thyroid is stimulated to produce thyroxine by the action of the pituitary gland at the base of the brain. This gland is sometimes referred to as a master gland because of the wide spectrum of hormones it controls. Prolactin, a hormone that in mammals is involved with the regulation of mother's milk, in salamanders serves to promote lar-

Figures 12-8A & 8B:
*A larval tiger sala-
mander (**A**) collected
from a temporary
pool near Crystal
Lakes, Colorado (a
mountain community
at about 8000 feet).
The tiger (**B**) after
transformation.*

val growth and inhibits transformational changes. The whole developmental sequence is probably regulated by a portion of the brain called the hypothalamus, which is in turn stimulated in some way by the animal's interaction with its environment.

Something as simple as a lack of iodine

Ambystoma tigrinum

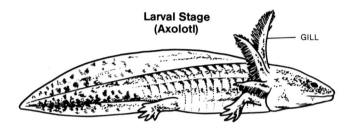

**Larval Stage
(Axolotl)**

GILL

Terrestrial ("Adult") Stage

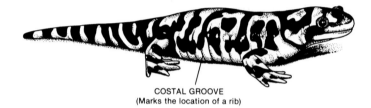

COSTAL GROOVE
(Marks the location of a rib)

Figure 12-9: Larval and young adult tiger salamanders are nearly the same size, but the latter has undergone substantial "remodeling" for the land life under the direction of the hormone thyroxine.

can short-circuit the whole process, however, because iodine is necessary for thyroxine to function. Artificial populations that never metamorphose have also been created by eliminating all individuals that transformed, so it's easy to see how nature may occasionally do the same thing when conditions dictate.

Watching the development of salamanders and other amphibians is one of the most fascinating aspects of this study. Because their eggs are large, hardy, and easily collected, they have been a favorite choice for many embryologists. It's proven relatively easy to graft portions of one embryo onto another, even from different species. Much has been learned about how early developmental processes are orchestrated. With a dissecting microscope or even a hand lens you can watch a ball of cells grow and differentiate into a complete, functioning animal. It's a sight worthy of anyone's wonder.

Finding the Ones That Didn't Fall In

Most amphibians, of course, won't be cooperative enough to fall into your window well. To find them you need to search lakes, ponds, and a variety of temporary pools and forest haunts, like rotting logs. The equipment needed is relatively simple. A small dip net, like those you get at pet stores to scoop fish from the aquarium, is useful. You will also need a long-handled net for snaring aquatic adults. These can be purchased or made with an old broom handle and a coat hanger to which a cloth cone has been securely sewn. A flashlight is necessary if you hunt at night, which is the best time for catching adults. If you can find or make a light attached to your headgear, so much the better, as your hand will be free to operate the net. Amphibians are relatively unperturbed by light, and you can often surprise them while mating and catch them easily.

Adult salamanders are fairly easy to keep in captivity. They don't eat much and are capable of going weeks without anything. Most amphibians require live food. They don't seem to consider things edible if they're not moving. Meal worms (the larvae of grain beetles) are available at pet stores. You can also get colonies of fruit flies from biological supply companies along with a container of "instant food" that will keep you in flies for a long time. These are useful as food for a variety of animals. If you get the shriveled wing mutants (referred to as apterous), you don't have to anesthetize them to handle

them successfully.

Keep the salamanders in cool water that is changed every two or three days, and allow them a spot to get out of the water. For short-term guests, you can put some water in an aquarium and wedge something under one edge so that the elevated end remains dry. For seasonal captives, you might want to make an outdoor vivarium (see page 208).

Frogs are sometimes prone to getting a fungus infection called redleg. It's apparently more apt to occur in captivity because the animals' bellies don't get sufficient aeration in metal cages. Changing the water often and allowing sufficient space for the frog to move around will help to prevent the condition. George W. Nace, a zoologist at the University of Michigan, has spent the last sixteen years learning how to care for amphibians so that scientists can maintain animals in the laboratory. His solution to redleg was to put green, rubber mesh material, like that used in grocery store meat counters, into his cages to provide proper ventilation.

Dr. Nace also struggled with the food problem. Frogs particularly like crickets, but eventually get rickets if fed nothing else. He provided tubifex worms to help balance the diet and dusted his crickets with a vitamin and mineral powder. He even got some frogs to eat sugar-coated

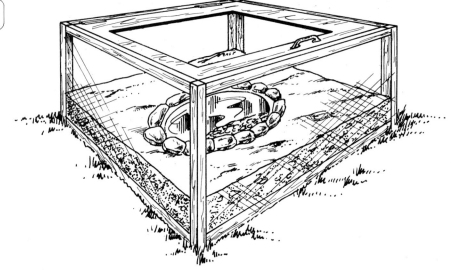

Figure 12-10: *A vivarium like the one shown will allow you to keep amphibians while not having to actively find food for them yourself. Release animals early enough for them to find shelter before winter sets in.*

Cheerios by vibrating the cereal on a platform, but you may not want to go to that extreme! Dr. Nace's company is called Amphitech, Inc. and is located in Ypsilanti, Michigan. Should you get particularly interested in amphibians, you can order adult and immature animals of a variety of species, as well as some of his specially designed equipment.

Small woodland salamanders can be kept successfully in a terrarium that is stocked with mosses, ferns and a piece of wood for cover. Other salamanders and frogs, however, wouldn't have sufficient space. If you have the time and interest, therefore, you might consider building an outdoor vivarium. Rucker Smyth in *Am-*

phibians and Their Ways describes a basic setup. Build a box from two-by-fours that is six or eight feet square and four feet high. Cover the entire enclosure with one-eighth inch mesh hardware cloth and fashion a door at the top. Fill the enclosure with a foot of dirt. Bury a large plastic dishpan to within an inch of its top. Place a fringe of stones around it to minimize the amount of dirt knocked into the water. Construct a ramp-type arrangement within the pan out of sand or rocks so that an animal can get out without too much trouble. Many insects will be able to enter through the mesh to be an unwitting food source. You can also place some rotting meat in the enclosure from time to time to attract flies. Lettuce left in a damp corner will attract slugs. Setting up a light at night will attract many insects. Animals can be placed in the vivarium for most of the season, then can be released before frost sets in. Avoid the temptation to over-crowd.

Caring for Eggs and Tads

Perhaps the most fascinating aspect of studying amphibians is watching their development from egg to tadpole or larva and from juvenile stages to adult. Collecting eggs and tads is relatively easy. To find the former, search in the spring after a wet spell. Look in the same locations as for adults, and check for black or dark-colored bodies encased in a jelly-like matrix. Sometimes eggs will be deposited in fairly large masses, and other times they will be single, in long strings or small clumps. Eggs may be found at the surface, attached to the stems of aquatic plants either at or below the water line, or attached to submerged objects.

To watch egg development, you'll find it necessary to collect the eggs and care for them inside. Here are some tips from Mr. Smyth:

- Avoid metal pans or other containers. Metal ions may be poisonous for the eggs.
- Use cold water that has not been chlorinated. If chlorinated tap water is all that is available, you can let it sit out uncovered overnight to let most of the chlorine bubble out.
- Give eggs plenty of room to hatch. If necessary, break up large egg masses into smaller units.
- Change the water at least every three days.
- Don't allow eggs to freeze. (However, water should not be warm. Cool water is best.)
- Eggs that normally float on the surface should be barely covered with water. If they sink to the bottom they will be less likely to hatch.
- Eggs collected from streams where there is normally a brisk flow of water will be unlikely to hatch unless they are aerated in some way, perhaps with an aquarium "bubbler."

For tadpoles, follow these guidelines:
- As with eggs, don't overcrowd.
- Change water frequently—every

two or three days.

•The tadpoles of frogs are vegetarian and can be fed lettuce or spinach leaves that have been boiled until soft (about a half an hour in an enamel pan to avoid metal ion contamination). Salamander larvae are harder to care for because they are carnivorous. If you keep them in an aquarium with live pond weeds and mud there will be plenty of small crustaceans, such as daphnia, for them to eat, but it is difficult to keep things clean and the water aerated. You can buy the water flea, daphnia, at most pet stores or try the ever-popular fruit fly. If you decide to build a vivarium, salamander larvae should do well there.

As you watch the transformation of salamander from a slender fish-like form to a four-legged land walker, you can't help but marvel at how a creature can be so completely remade by the interaction of environment, body chemistry, and genetics. It's fascinating, too, how several major groups of animals—crustaceans, arachnids, insects, and vertebrates—faced with similar problems in exploiting dry land and the plants that first conquered it, found such different solutions using the "equipment" that eons of selection had given them. At any rate, I'm looking forward to the next tiger that drops by.

REFERENCES

Alper, Joseph. 1985. Frog Factory. *Natural History*, vol. 6, no. 4 (May).

Bishop, Sherman C. 1967. *Handbook of Salamanders*. Ithaca, New York: Cornell University Press.

Borgens, Richard B. 1977. Skin Batteries and Limb Regeneration. *Natural History*, vol. 86, no. 8 (October).

Margulis, Lynn and Karlene V. Schwartz. 1982. *Five Kingdoms: An Illustrated Guide to the Phyla of Life on Earth*. San Francisco: W.H. Freeman and Co.

McLoughlin, John C. 1980. *Synapsida, A New Look Into the Origin of Mammals*. New York: The Viking Press.

Morholt, Evelyn, Paul F. Brandwein and Alexander Joseph. 1966. *A Sourcebook for the Biological Sciences*, 2nd ed. New York: Harcourt Brace Jovanovich.

Owen, Ellis. 1975. *Prehistoric Animals*. London: Octopus Books Ltd.

Sherman, Cynthia Kagarise and Martin L. Morton. 1984. The Toad That Stays on Its Toes. *Natural History*, vol. 93, no. 3 (March).

Smyth, H. Rucker. 1962. *Amphibians and Their Ways*. New York: The Macmillan Company. A good book for the naturalist; it offers many hints for field studies.

Twitty, Victor Chandler. 1966. *Of Scientists and Salamanders*. San Francisco: Freeman and Company. An entertaining tour of amphibian embryology and biology as it related to the author during his career.

Whitfield, Philip, ed. 1984. *Macmillan Illustrated Animal Encyclopedia*. New York: Macmillan Publishing Company.

Wood, S.P., A.L. Panchen and T.R. Smithson. 1985. A Terrestrial Fauna From the Scottish Lower Carboniferous. *Nature*, vol. 314 (March 28). Describes beautiful new fossils of Carboniferous age, including the earliest, well-preserved skeleton of an amphibian.

Young, J.Z. 1981. *The Life of Vertebrates*, 3rd ed. Oxford: Clarendon Press. This textbook on vertebrates covers anatomy, biology, and paleontology quite thoroughly.

13

Dinosaurs in the Garden

The dinosaurs were quite brave. They strutted over the light dusting of snow on the sidewalk and approached the food. They cast darting glances from side to side, sometimes observing the black, four-legged mammals that lay on the grass and other times eyeing suspiciously the two-legged mammal that stood, partially concealed, behind a shadowy screen. With quick motions the dinosaurs captured morsels of dog food in their beaks and flew away.

The particular dinosaurs I speak of are commonly referred to as starlings. Their more impressive cousins, such as *Brontosaurus* and *Tyrannosaurus rex*, died out 65 million years ago in a cataclysm that was so pervasive and nearly complete that dinosaurs have become a metaphor for failure. Actually they were a highly successful group that persisted for 160 million years (225 million, if we count birds).

At first blush, birds seem to have little in common with our standard image of dinosaurs. Birds are warm-blooded creatures with a unique covering called feathers, and they share with mammals a large brain relative to their body weight. Dinosaurs have been portrayed as cold-blooded reptiles of gargantuan proportions that lumbered about in tropical habitats, their massive limbs coordinated by barely adequate brains. This image is now known to be incomplete, and in

some cases, just plain wrong. Since the mid-1970s many of our views of dinosaurs and their world have changed. Before we look at the opportunistic starling as an example of a successful modern dinosaur, let's check out his reptilian lineage.

Archaeopteryx: The "Bird" That Couldn't Fly

In 1861 a beautiful fossil was found in some fine-grained shale in Bavaria. Dr. Karl Haberlein traded his medical services to men in the quarry for some of the fossils they unearthed. The British museum paid 700£ for this particular fossil, later dubbed *Archaeopteryx lithographica*. Dr. Haberlein acquired a dowry for his daughter, and posterity acquired an exceptionally clear record of an unusual animal. Two other *Archaeopteryx* were eventually discovered in 1877 and 1956. All were beautifully preserved crow-sized creatures that were very reptile-like in appearance, but it was obvious that they were covered with feathers very similar to those of modern birds.

In 1970 Dr. John H. Ostrom was studying other fossils that had been discovered in the same quarry but had been classified as pterosaurs—one of the small flying dinosaurs that also lived in the Jurassic

Figure 13-1: *A feathered, but flightless* Archaeopteryx *chases an insect through the Jurassic underbrush, startling a small pterosaur.* Archaeopteryx *was approximately chicken size. Pterosaurs came in all sizes from that of a sparrow to those possessing 40 foot wingspans.*

period. He found impressions of feathers that others had overlooked, and the fossil (originally unearthed in 1855) was recognized as another *Archaeopteryx.* The "last" *Archaeopteryx* was actually found in 1951, but was classified as a small Coelurosaurian dinosaur until Dr. Ostrom observed traces of feathers when examining the fossil with oblique lighting techniques in 1973.

Archaeopteryx's mix of characters was confusing. The skull resembled that of many other small dinosaurs, featuring a pair of jaws lined with teeth. The space that housed the brain was larger than most reptiles, but there was no evidence of the enlarged cerebrum and cerebellum found in birds. The vertebral column, including a long, tapering tail, was distinctly reptilian, but the tail was adorned with parallel rows of feathers. The hands had three long fingers tipped with claws, but both the hands and arms were feathered. *Archaeopteryx* possessed the fused collar bones of birds while lacking the heavy keel bone needed for the attachment of flight muscles. The bones themselves were heavier than those of birds, as evidenced by the lack of air spaces within them.[1]

The implications from the physical information is that *Archaeopteryx* was probably not capable of powered flight, but may have been able to glide. Like other ground-dwelling dinosaurs of that size, *Archaeopteryx* most likely fed on insects under the canopy of a coniferous

---

[1]In 1986 two crow-sized, 225-million-year-old fossils were found near Post, Texas, and given the name *Protoavis* because of their bird-like characteristics. The skulls of the fossils were nearly identical to modern birds except for teeth in the front. *Protoavis* also possessed a breastbone and flight-modified limbs. In addition to its dinosaurian-like teeth it sported clawed fingers and a tail. *Protoavis* was probably closer to being the ancestor of birds than *Archaeopteryx*, but still displays a strong relationship to dinosaurs.

Bird or Dinosaur?

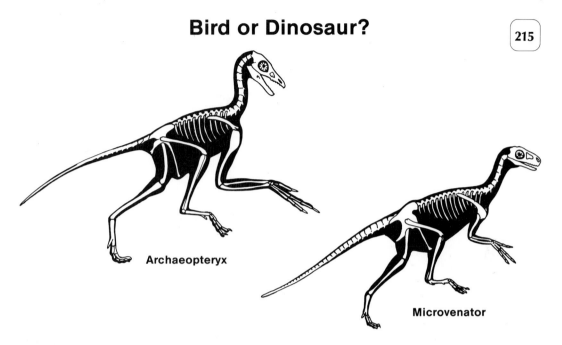

Archaeopteryx

Microvenator

Figure 13-2: John Ostrom of Yale demonstrated that the joint anatomy of Archaeopteryx and small dinosaurs like Microvenator are virtually identical. Without a clear indication of feathers on some fossils it's easy to see how early researchers were confused. There is now little doubt of a close relationship between these small, bipedal dinosaurs and birds. (Redrawn from Bakker, 1975)

forest. And the feathers? As with birds, they were very likely an effective insulation and imply that the animal they covered was warm-blooded.

Hot Blood, Cold Blood

What does it mean to be a warm-blooded creature? Human beings, like birds, are warm-blooded, but it doesn't necessarily mean that your blood is hotter than that of a lizard sunning himself on a rock. It may even be colder. It does mean, however, that you can maintain a relatively high and constant internal temperature, regardless of what the weather is up to. The advantage to the animal is that he can stay alert and active in a much wider range of conditions. Warm-bloodedness, also called endothermy, is a major innovation. There is a price for an endothermic animal to pay, however, for it takes more food energy to fuel such a creature.

Birds and mammals were once thought to be the only endotherms. Dinosaurs, with their reptile-like appearance, were

Figure 13-3: Small birds, like this pine siskin, pay the price for warm bloodedness: they must forage constantly, especially in cold weather, to maintain their body temperature. These birds were captured, banded and released by a licensed bird bander. It's illegal to trap or keep wild birds (although starlings are excluded because of their "pest" status).

Bird Bone Structure

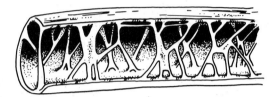

Figure 13-4: Bird bones are light in weight yet strong, an admirable adaptation for flight. They are filled with air spaces that are often continuous with a system of air sacs in a bird's body. Strength is provided by the lacework of struts that partition the bone interior.

more or less assumed to be cold-blooded ectotherms like modern reptiles. Much evidence now implies otherwise.

It would seem unlikely that you could tell anything about the physiology of an animal that has been dead a hundred million years or more, but an animal's energy strategy does leave its mark—both on the animal itself and on the community of creatures to which it belongs. For example, the bones of endotherms and ectotherms are distinct. The bones of warm-blooded animals are more richly filled with blood vessels. The spaces these vessels tunnel out within the bone are called Haversian canals. The bones of cold-blooded animals, lacking the high concentration of canals, are more compact. In addition, the bones of these latter creatures often show growth rings due to the reduced rates of growth during periods of cold or drought. According to some authorities, dinosaur bones display a warm-blooded pattern of growth.

Climate changes over time can also reveal something about an animal's physiology. Small ectothermic animals like lizards and turtles can hibernate in temperate zones during winter, but their larger cousins cannot. Large ectotherms are restricted to mild climates because of their inability to maintain adequate body temperature over a sustained period of cold weather. Endotherms, with adequate food and body insulation of hair or feathers, can survive nicely in temperate and arctic conditions. Although the Jurassic period, during which dinosaurs were prominent, displayed very mild temperature gradients, the succeeding Cretaceous period sported rhinoceros-sized dinosaurs that lived within the arctic circle of that time. There were also large marine forms in the far north.

Some of the most interesting evidence for warm-blooded dinosaurs is revealed in "petrified ecosystems": specifically, the ratios of dinosaur predators to their prey. Since warm-blooded animals need to eat more to maintain their internal fires, the same amount of prey species will support fewer endothermic than ectothermic predators. In modern ecosystems, for example, ectotherm predators account for 40 percent of animal species, whereas endotherm predators constitute only 1 to 3 percent of animals in their ecosystems. Predator-to-prey ratios of dinosaur communities from the Triassic, Jurassic, and Cretaceous all fall in the 1 to 3 percent range.

The presence of body insulation is a strong indicator of endothermy. Small animals, with their large surface area to volume ratio, lose heat especially fast— just as a cup of coffee cools off much more quickly than a kettle filled with water. Insulation can thus play an important role in heat regulation, and it was used by at least one prehistoric species—a Jurassic pterosaur that had dense covering of hair or hairlike feathers. Pterosaurs were apparently not the leathery skinned beasts often portrayed in cartoons. They may have resembled a cross between a long-billed bird and a bat.

Learning to Fly

It would seem that the distant cousins of my backyard starlings looked like birds long before they acted like birds—that is, before they could fly. What does it take to be a successful flyer? One needs wings, large muscles anchored to strong bones to power those wings, a light but strong framework, good brain power and coordination, and a high metabolic rate to keep those wings moving. Among vertebrates, this combination was achieved three times: the pterosaurs, birds, and bats.

Pterosaurs were a successful experiment in flying. The earliest types date back 200 million years, and the later varieties, the famous pterodactyls, lasted until the great Cretaceous extinctions. Pterosaurs came in sizes that ranged from the di-

Three Ways to Fly

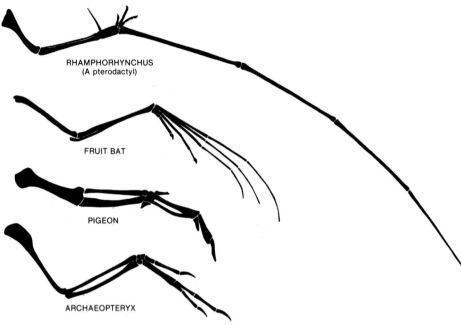

RHAMPHORHYNCHUS
(A pterodactyl)

FRUIT BAT

PIGEON

ARCHAEOPTERYX

Figure 13-5: *Flight evolved in three distinct ways among vertebrates, all involving the bones of the forearm. The first fliers, the pterosaurs, formed wings from a greatly extended fourth digit. Birds utilize the whole forearm, with some bones being fused. Archaeopteryx clearly shows the same pattern as birds, but lack of a keel bone for attachment of flight muscles implies that the latter was not an active flier. Bats, the last vertebrates to exploit the air, use all the bones of the hand for a wing surface.*

mensions of a sparrow to those of a jet plane. *Quetzalcoatlus northropi*, with a wing span of 35 to 40 feet, was found in Texas (of course). The pterosaur, bird, and bat designs are, however, all a bit different and imply separate lines of descent. In pterosaurs the wing surface was created by a membrane which stretched from a greatly elongated fourth finger to the animal's body. In birds, the wing is formed from the entire arm; in bats, wing design incorporates arms, legs, and elongated fingers on the hands.

The structure of these flying vertebrates provides some clues as to how flying was achieved in each case. The two main theories are usually referred to as the "trees down" and "ground up" approaches. It would seem logical that an animal might acquire powered flight by first utilizing flaps of skin to glide from upper to lower branches in the canopy. If this proved beneficial, natural selection would favor more elaborate membranes, and eventually, movement of the arms—perhaps to stabilize the glide—would lead to powered flight. In the case of pterosaurs and birds, however, the "ground up" technique now seems more likely. *Archaeopteryx* had strong running legs, much like its Coelurosaur cousins, and was most likely a ground-dwelling insect eater. Its long, feathered arms were probably used to balance the animal as it lunged at prey with its toothed beak. Like insects, when the "wings" reached some critical size they displayed aerodynamic properties which may have helped *Archaeopteryx* escape from predators.

Pterosaurs also had feet that were capable of walking, and although no one has found such convenient transitional creatures as *Archeopteryx*, there is a good chance that pterosaurs started out on the ground. Bats, on the other hand, are helpless on the ground. They may have been tree-dwellers with a habit of hanging upside down. This would free both the arms and legs to respond to nature's challenges by developing into an airfoil. Fifty million years ago bats found a successful niche as nocturnal, flying hunters.

Bats or birds might never have found their niches, however, if the pterosaurs had not died out with the dinosaurs 65 million years ago. For unknown reasons birds did survive. Both they and the mammals inherited a virgin world into which they spread and diversified. One of the very successful orders of birds have been the *Passeriformes*, or perching birds. Your backyard starling is a sterling example of this group.

History of a Pest

Contrary to what many people might believe, starlings were not designed specifically to confound the human race—even though their fossil record does start about the same time in the Pleistocene. They probably have lived near human activity since man became a farmer 10,000 years ago. Starlings belong to a family of birds, the Sturnidae, that includes over a hundred species. Members of this group are small to medium-sized birds with strong legs and a bill that is generally stout and strong. Wings are long and built for fast flying. The sexes are usually similar in appearance. Juvenile birds may be rather dull in coloration, while adults often sport glossy, iridescent feathers.

Starlings get into trouble with people because, like people, they are hardy opportunists that are smart, aggressive, and eat anything. The European starling, *Sturnis vulgaris*, ranged over Europe, Africa, and Southeast Asia for most of the time

A Bird Body Plan

**The Common Starling
(*Sturnis vulgaris*)
In Its Fall Plumage**

Figure 13-6: In the fall, the wing tips of starling feathers are white, giving the birds a speckled appearance. This starling in flight also demonstrates some of the features that have made starlings so successful: a strong beak useful for prying up grubs in grassland, strong wings adapted for agile flight and stout walking legs.

humans have been aware of them, but they now have nearly worldwide distribution. Starlings were transplanted to various countries in the last century in the hope that their insect-eating habits would be a boon to agriculture. Several attempts to found populations in the United States failed in New Jersey, Massachusetts, and Oregon before Eugene Scheifflin, in an attempt to import all the birds mentioned by Shakespeare in his plays, successfully started a population in Central Park with a hundred birds released in 1890 and 1891. The first starlings to breed in the U.S. may have nested, perhaps fittingly, under the eaves of the American Museum of Natural History.

Within 55 years of Mr. Scheifflin's tribute to the Bard, starlings had spread to British Columbia. In the 1950's they were found in Alaska. They now reside in all fifty states. During the time of their range expansion, however, their insect consumption has not compensated many farmers and feedlot operators for loss of grain and the spread of certain diseases. The starling's habit of feeding and roosting in large flocks makes for noisy neighbors, and even "bird people" resent the fact that the starling outcompetes many of the native species.

I have to admit, however, to a certain admiration for this brassy bird that mimics Man in so many ways—including his speech. Mynah birds belong to the same family, so it's not too surprising that starlings can imitate a variety of sounds. In the 5th century B.C. the European starling was kept as a cage bird in Greek houses. Pliny describes a bird that could speak Greek and Latin "and moreover practiced diligently and spoke new phrases every day, in still longer sentences." More recently Margarete Sigl Corbo in a book called *Arnie the Darling Starling* describes her experiences raising a starling. Her grandsons spoke to it constantly, and one day it started talking back, even using phrases at seemingly appropriate times. In nature it's not clear how this talent for mimicry benefits the starling, but adult birds do incorporate parts of the songs of other birds that live nearby. In part, the length and complexity of a male's song may clue the female in to the fact that he is older and more experienced than a bird with the "basic" song.

Whether or not you attempt to raise a starling (and these are one of the few birds you can capture without a permit), you can observe much about this animal's behavior with a little patience and a pair of binoculars.

A Starling Profile

Sometimes the starling is mistaken for more than one kind of bird because their appearance changes significantly with age, sex, and the time of year. Newly hatched birds are a mouse-brown with dark bills until the fall molt. When the new feathers come in they are tipped with white, which gives the animals a speckled appearance. Males have the smallest "spots," and they disappear more quickly during the winter so that by breeding time in the spring they are nearly gone. Adult females, first-year males, and first-year females appear progressively more spotted. As breeding time approaches, a starling's legs become redder and the bill changes from dark brown to yellow. A male's bill becomes a brighter yellow than the female's and is steel-blue at the base, whereas the base of a female's bill is pink. Another clue for sex determination comes from the eyes. The male's eyes tend to be all dark, while those of the female have a pale ring around the edge. An adult male's throat feathers are also longer and glossier. The adult animal struts around with a predominantly black cloak of feathers, shimmering with green and purple iridescence, particularly around the neck.

The typical starling weighs 75 to 100 g, is 20 cm long, and walks, rather than hops, around on large feet that span 50 mm.

Sexual Differences in Starlings

Male

Female

Figure 13-7: You can often tell males from females by looking at the eyes. Females have a pale ring at the margins of the eye, but males do not. At the end of a season males may be partially "bald" as head feathers abrade away. A male's white "spots" wear away more quickly because they have less white on the tips of their feathers. Females have a larger "brood patch" on their rumps as they perform most of the egg incubation.

Each toe has a sharp claw. The bill is 25 mm long, strong, pointed, and suited for a diet of fruit, seeds, and insects, as well as the dog food near my back door. The prying behavior that starlings and other members of their family use in digging dormant insects out of the soil is an important adaptation for feeding in temperate grasslands. Tick birds, members of a sub-family of the Sturnidae, use similar

Figure 13-8: Starlings in winter plumage

techniques in digging lunch from the furrows of skin on large grazing animals.

Starling Behavior

Next to looking at each other, people seem to enjoy watching birds. I never was a particular bird fan while growing up, but have found their behavior quite fascinating as I've shared my children's interest in bird watching. Most people see birds at a feeder. There you get plenty of opportunity to observe aggressive and dominance-related behavior. Observing starlings around their nesting holes in the spring will reward you with activities designed to attract mates, establish territories, and care for the young. Careful observations over time may tell you something about starling migration patterns in your area and how starlings extend their ranges. Starling roosting behavior may remind you of a film clip from Alfred Hitchcock's "The Birds."

Aggression

We had an early taste of winter this year. Snow blew in to coat the ground for a few days in the middle of September. Starlings gathered on power lines and roof eaves near the feeder, and their personalities were exposed for all to see. As they lined up on the power line one bird threatened another with fluffed out feathers and an impassive stare. The staring animal held its head high and raised its crown feathers. The other animal adopted a submissive crouch. The aggressor stabbed a bit at the other with its beak, however, and the second bird began bill wiping, which involved rapidly wiping its bill on either side of the perch. This behavior during feeding usually signifies that the animal is finished eating and may tell other birds that it is no longer a competitor. In aggressive encounters it seems to be adopted as a form of appeasement. The first bird on the power line relented, and they sat side by side for a while.

Two birds at the feeder got into a scrap and engaged in a "fly up." Both birds flew up to a height of 1 to 2 m, stabbing with their bills and kicking with their feet. They landed amid squawks and chattering, including a rapidly repeated "chacker-chackerchacker" sound.

Another bird lowered its head and charged an opponent with its beak open. The second bird flew away. An arriving animal settled to the ground, rapidly flicking the tips of its extended wings. This wing-flicking behavior is also used by birds arriving or leaving a roost in conjunction with "mobbing calls" that elicit help from other starlings when a predator is near. In feeding behavior the wing flick is an aggressive signal.

Back on the power lines various birds were practicing their sidling behavior. The aggressor sidles toward another bird, forcing it along the wire until it reaches some obstacle—like another starling. On a branch, one bird may literally walk another off the branch.

Courting and Domestic Behavior

Males pick out suitable nesting holes 10 to 30 ft. off the ground in buildings or trees. Hole diameters need to be an inch and a half or larger. Once a site is selected males will defend it by using many of the aggressive strategies already mentioned: fluffing, bill wiping, wing flicking, and sidling.

The next step involves advertising for a

GENERAL SONG *High pitched whistles, rattles and clicks audible at close range. Variable in content and may imitate other birds' songs.*

> *Sex: Either*
> *Season: Spring, summer, fall*

STARING *Upright stance, crown feathers raised while staring at opponent. The bird on the right, with head higher and bill open, is more apprehensive.*

WING WAVE *While perched, bird spreads wings and rotates them. A vacillating, high-pitched squeal call often accompanies this behavior.*

> *Sex: Male*
> *Season: Spring, summer, fall*
> *Associated with: Courting*

WING FLICK *Tips of wings extended and rapidly flicked. May be associated with squeal call or "mobbing call."*

> *Sex: Either*
> *Season: All*
> *Associated with: Aggression at feeding and arriving at flock*

CROWING

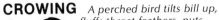

A perched bird tilts bill up, fluffs throat feathers, puts tail down and vertical with bill closed. Continuous chortling call, variable and unmusical.

Sex: Male
Season: Spring, fall, winter
Associated with: Territory
(Defense of nest site)

SIDLING

Bird moves sideways along branch, forcing other bird off.

Sex: Usually male
Season: All
Associated with: Territory
(Defense of nest site)

FLUFFING

Feathers are fluffed out while facing another bird.

Sex: Either
Season: All
Associated with: Aggressive encounters

BILL WIPE

Bird rapidly and repeatedly wipes bill on either side of branch. Submissive display.

Sex: Either
Season: All
Associated with: Feeding and aggression

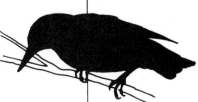

CROUCH *Body is lowered and feathers sleeked in a submissive gesture.*
Sex: Either
Season: All
Associated with: Aggressive encounters

FLY–UPS *Birds fly up, calling at each other while stabbing with beaks and kicking.*
Sex: Either
Season: All
Associated with: Aggression during feeding

FLOCK CALL *Short "djjj" sound.*
Sex: Either sex of juvenile birds
Season: Summer and Fall
Associated with: Feeding in large groups

mate. A male will give a squeal call and wave its wings in a rotating fashion when any starling passes overhead. Most don't respond, but if an unmated female passes by, she may approach the nest. Males may then begin dragging twigs and bits of leaves into and out of the nest in a rather blatant invitation. This may go on for some time, interrupted by feeding and other activities. When birds have paired, however, they become inseparable and do everything together. Eventually the female takes over nest building, often throwing out whatever the male has already collected and lining her home with dead leaves and grasses.

About the time of pairing you may see chase flights between the couple. Mating occurs about this time, also, and the only preamble may be the female pecking a bit at the male's neck. When the male mounts the female, he often waves his wings to keep his balance. Some scientists have speculated that this may be the origin of the wing-waving display.

Look for nest sites any time of the year except August, when the birds molt. Check out older trees near the edge of clearings and various nooks and crannies around buildings. Watch for males performing their wing waves and listen for their crowing calls. Sometimes eggs may be found on the ground near the nest, especially if the nest has been disturbed. Starling eggs are blue and look very similar to robin's eggs. Starlings usually lay a clutch of four to five eggs that are incubated by both male and female. Females have larger brooding patches on their rumps, however, and account for most of the warming. Incubation takes 12 days.

The nestling phase lasts 23 days. Chicks begin begging calls when they are one to two days old. In addition to their distant call, they have one for when the parents are close that sounds rather like "cheer-cheer-cheer." Males may start more than one family, but usually spend most of their time near the first brood. The number and size of clutches depends largely on the quantity and quality of the food supply.

You may hear distress calls from juveniles. These are high-pitched, raucous, penetrating squeals. Adults use the calls less often, but the calls increase in frequency during molting. Such calls can disperse entire roosts in winter. People have tape recorded them and used them as "scare calls" in fields.

Roosting

Perhaps nothing has gotten starlings in more trouble with people than their habit of roosting in large flocks. They are noisy, messy aggregations that may rise up like one organism and swoop down like locusts onto a field. Roosting is not peculiar to starlings, but starlings are less fussy than most birds in where they roost. You can find them in cattail, reed, and alder stands; in pine, cedar, maple, and mixed deciduous stands; and under bridges, on buildings, and in bell towers. Some roosts may have as many as a quarter million birds in them!

Roosting sites may be quite persistently used. In England, one study demonstrated that 70 percent of roosting sites were in use more than five years and 12 percent were used for over 50 years. Other records showed that some sites were in use continuously for over a hundred years. It would seem that there might be serious disadvantages to this behavior. Predators would certainly know where to go for a snack, and all the birds would be working over roughly the same territory for food.

The other side of the situation, however, is that there is both safety in numbers and information exchange regarding food supply that makes the behavior worthwhile. Different birds in the flock may reap different benefits from roosting. In a large flock, the chance of any one bird being taken by a predator is small, and the chance of some bird sounding an alarm is very high. Older, more experienced birds may benefit most from this because they get the best spots, away from the periphery. Juvenile birds, on the other hand, gain more benefit from knowing where the best sites for feeding are. All birds gain protection from predators during feeding because there are more eyes to see the danger.

Try this experiment with starlings: make decoy birds in two positions. In one position, the beak of your imitation animal should be touching the ground as if it were feeding; in the other, have the decoy upright and "alert." The decoys need not be too elaborate. They should be basically black and the right size, sporting a yellow bill. Try painted cardboard shapes with stiff wire legs you can poke into the ground. Use one decoy first and record the number of starlings that land in the yard in a given amount of time—say twenty minutes. Repeat with the other decoy. You should find that your "feeding decoy" will attract more animals.

Starlings on the Move

Starlings provided scientists with a good model for how animals may extend their ranges. Starlings were observed to build up a population in an area, then begin fall and winter excursions into neighboring regions. Within five years of these explorations, breeding pairs became established. Six years after being introduced to Central Park, starlings colonized Long Island, Brooklyn, and nearby suburbs. By 1915 they were residents of Halifax, Nova Scotia, and two years later some were found in Savanna, Georgia. In another dozen years their range was considered to be anywhere east of the Mississippi River. As I've already mentioned, the Great Plains and Rocky Mountains also proved to be inadequate barriers, and starlings can now be found throughout the U.S. For other such inadvertent "experiments" refer to George Laycock's book, *The Alien Animals.*

The Joys of Dinosaur Watching

Of course, there's no reason to limit yourself to observing only one kind of dinosaur. A wide range of colorful and noisy examples will visit any feeder. But don't overlook the "pests" and "plain janes." The fact that certain species are common means they have successfully adapted to a world full of people—and that's no mean feat. I've listed one or two bird identification keys at the end of the chapter, but it would be worth your while to check out the two-volume *Stokes Nature Guide to Bird Behavior*, which gives you more than just physical descriptions of the animals.

I can't imagine a world without birds and sometimes wish a few other dinosaurs had survived the Cretaceous "fall." (Perhaps a similar sentiment motivates others to search out the Loch Ness monsters of the world.) Nevertheless, I'm agreeable to the way things turned out. Otherwise, a dinosaur might have written this book, and the order of this chapter and the next would undoubtedly be reversed.

REFERENCES

Bakker, Robert T. 1975. Dinosaur Renaissance. *Scientific American*, vol. 232, no. 4 (April).

Burton, Jane and Dixon Dougal. 1984. *Time Exposure—A Photographic Record of the Dinosaur Age.* New York: Beaufort Books. An interesting little book with color photographs that help recreate the dinosaur's world.

Desmond, Adrian J. 1976. *The Hot-Blooded Dinosaurs.* New York: Dial Press/James Wade. One of the first texts to outline the arguments for homeothermic metabolism in dinosaurs.

Feare, Christopher. 1984. *The Starling.* New York: Oxford University Press. The major single-volume reference on starlings.

Greenberg, Joel, ed. 1986. Oldest Bird and Longest Dinosaur. *Science News*, vol. 130, no. 7 (August 16).

Laycock, George. 1966. *The Alien Animals.* Garden City, New York: Natural History Press.

Lewin, Roger. 1982. *Thread of Life.* New York: W.W. Norton and Co.

McLoughlin, John C. 1979. *Archosauria, A New Look at the Old Dinosaur.* New York: Viking Press.

Peterson, Roger Tory. 1963. *The Birds.* New York: Time Inc.

Robbins, Chandler S., Bertel Bruun and Herbert S. Zim. 1966. *A Guide to Field Identification, Birds of North America.* New York: Golden Press.

Simpson, George Gaylord. 1983. *Fossils and the History of Life.* New York: Scientific American Books. An excellent reference.

Stokes, Donald W. 1979. *A Guide to Bird Behavior,* Vol. I. Boston: Little, Brown and Co. This behavior series is very good for naturalists. So far there are two volumes on birds.

Tweedie, Michael. 1977. *The World of Dinosaurs.* New York: William Morrow and Co. A large-format book with many excellent color paintings.

Weatherhead, Patrick J. 1985. The Birds Communal Connection. *Natural History*, vol. 94, no. 2 (February).

Young, J.Z. 1981. *The Life of Vertebrates*, 3rd ed. Oxford: Clarendon Press.

14

Of Mice and Men (and Dogs and Cats)

Those who study natural history end up with mementos on their shelves that others may find strange. I have a collection of skulls, for example, of such creatures as squirrels, mice, sheep, rabbits, and a friend's prize boar. Through the acquaintance of a physical anthropologist I've acquired casts of various human skulls, from African "pre-men" like *Australopithecus* to homicide victims of the purportedly advanced *Homo sapiens*. The skulls of the mice fit nicely within the eye sockets of those of the men, yet both are mammals. Like the dinosaurs, mammals have taken advantage of many opportunities to fill ecological niches, and this has resulted in vast differences in mammal sizes and shapes.

Four mammals dominate the ecology of my backyard: dogs, cats, mice, and people. Chances are they are important in yours, too. Though these animals are not necessarily a representative sampling of mammals, a brief look at their history and biology provides a decent introduction to the group. In addition, all four have histories that are curiously intertwined.

Mice and men might seem to have the least in common, but they both exemplify biological success stories that are quite amazing. Mice and their other rodent cousins account for half of all living species of mammals. Men, on the other hand, have acquired importance far beyond their numbers by exploiting a tendency to "braininess" that has led to self-awareness and abstract thinking. Such abilities generate both power over nature and the necessity to control that power or suffer the consequences.

Mice have been successful in spite of men. Of course, man's agricultural revolution certainly didn't hurt rodents, who quickly learned to set up housekeeping in granaries. Yet mice and their relatives would undoubtedly be around even without us.

Other animals survive because they are useful to us: chickens, pigs, cows, and many other domesticated creatures thrive, while the wild stocks from which they originally arose have suffered extinction. Many people, in fact, only measure an animal's or plant's worth by its direct effect on people. That, as I hope you come to appreciate with your own study of nature, is a dangerous attitude.

A few species of animals have something approaching a partnership with man. Dogs and cats fit into this latter category. The dog-man relationship goes far back, perhaps to a point when *Homo sapiens* was new to the world. Our life styles and affections have meshed well. The domestication of cats is historically documented and gives insights into how interdependencies among creatures get started. From mice we can see something

Figure 14-1: *Most people encounter nature in the form of pets and pests. Pets give us love and company and pests live off our excesses, but a host of other plants and animals are necessary for our continued survival and well-being.*

of the early mammals that held their own in a world dominated by dinosaurs.

Surveying the Local Rodent Population

You may find it interesting to observe some of the mice and other small rodents that share your backyard. If mice are pests in the house or near trash containers, you may have collected a small sampling in conventional mouse traps. Systematic trapping, with traps that capture the animals alive, will allow you to see a greater variety of small rodents and the kinds of habitats they prefer. Before attempting trapping of any kind, however, make sure to read "Notes for the Prospective Naturalist" in Chapter 1.

Perhaps the most common live traps in the United States are Havahart traps. They are baited with food appropriate to the kind of animal you wish to catch and are designed to allow the animal to enter far enough into the trap to allow the door to shut and lock. This feature not only keeps the trapped animal inside, but keeps predators away from the prisoner. Set out such traps late in the day and check them early in the morning so that trapped animals don't get baked by the daytime sun.

A simpler trap you can make yourself is described by Gerald Durrell in *A Practical Guide for the Amateur Naturalist.* Cut the lid off a large coffee can to fit just inside the top rim. Pierce the coffee can just below the rim at either end of a diameter line. Pass a fairly stiff wire through these

Figure 14-2: *Mus musculus, the common house mouse, has found it profitable, by and large, to take up residence near human beings. This individual, however, had to be rescued from another house guest—the domestic cat.*

They avoid large open spaces when they can and prefer to move along walls or the weedy margins of gardens. Traps can be baited with the traditional cheese, but fresh greens and grain work just as well, as does a mixture of peanut butter and rolled oats, mixed so that the oats are held together but the peanut butter is no longer sticky. If you lay out a number of traps in a gridwork pattern that encompasses several habitats, such as open field, fencerow, bank and ditch, you may find a greater variety of rodents than you ever expected in your neighborhood. Observe these fascinating animals for a short time, providing adequate food and living space, and then release them near the point of capture.

Mice, rats, voles, and lemmings belong to the family *Cricetidae*. Here are a few small neighbors you're apt to meet in your backyard.

holes. Hinge the lid on the wire at its balance point so that the lid will lie flat within the rim when at rest. Bury the can up to the rim at a promising collecting site. Cover the lid with loose soil, twigs and leaves. Any rodent that steps on the edge of the lid will fall to the bottom. The drawback of this type of trap is that more than one animal may fall in and one or both could be injured. This sort of trap should be checked often.

The art of trapping involves thinking like the trapee. Therefore, think like a mouse, if that is what you seek. Mice can be expected to visit their nest holes and often do so along well-worn paths or "runways".

The **house mouse** (*Mus musculus*) followed us from Europe. As the name implies, he spends most of his time indoors, although in the summer he may go outside and join the deer mice. House mice are brown or gray-brown, shading to a lighter brown underneath. Their ears are nearly naked, and they have a long tail with close-lying fur. In some places they may be confused with harvest mice, but the latter have grooves on their front teeth

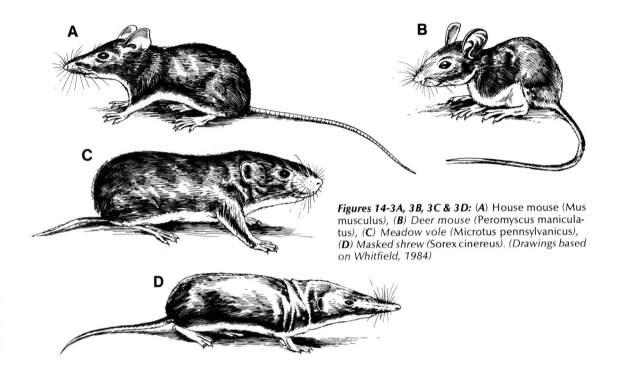

Figures 14-3A, 3B, 3C & 3D: (**A**) House mouse (Mus musculus), (**B**) Deer mouse (Peromyscus maniculatus), (**C**) Meadow vole (Microtus pennsylvanicus), (**D**) Masked shrew (Sorex cinereus). (Drawings based on Whitfield, 1984)

(incisors). Their total length (from tip of nose to tip of tail) is 150 to 180 mm (6 to 7 inches). They weigh 12 to 23 grams (.4 to .8 oz). House mice dine on grains, vegetables, meat, paste, glue, soap, and a variety of stored items.

Deer mice (*Peromyscus* species) are also called white-footed mice. Their white undersides distinguish them from house mice, whose belly fur always has a tinge of yellow. Their fur varies from gray or sandy to deep or golden brown. In cool woodlands they are gray, in the southeast they are variable in color, and in open and arid terrains they tend to be pale. All the young are grayish in color. Their tails are as long or longer than their bodies and have short hair. They are 141 to 195 mm (5.6 to 7.7 inches) long and weigh 12 to 31 grams (.4 to 1.1 oz.). In the fall, a few deer mice may try to join their house mouse cousins inside.

Voles (*Microtus* species) are also called meadow or field mice. They have short ears and a stocky body with "beady" black eyes. Their tails are shorter, relative to body length, than either deer mice or house mice—a useful characteristic for identification. They are usually brown or dark brown in color, but grayish or buff varieties are found, too. These mice prefer grassy or sedgy places in low vegetation.

The common vole is 120 to 188 mm (4.7 to 7.4 inches) long and the weight is 20 to 68 grams (.7 to 2.4 oz).

Shrews belong to the family *Soricidae* and occasionally turn up in mouse traps. The common shrew (*Sorex cinereus*) is grayish-brown, paler underneath. They may have a poorly defined stripe down the middle of their back and gray sides. Shrews are the shortest of mammals and closest to the early mammals that shared the world with dinosaurs. They have pointed noses and their tails are about half the total body length of 80 to 109 mm (3 to 4 inches). They are most easily distinguished from mice by their teeth. They have a continuous row of small, sharp-pointed, brown-tipped teeth from front to back, whereas mice have the two large front incisors with the grinding cheek teeth separated by a gap. Shrews operate at high velocity. Their hearts beat 800 times a minute (compare with a hummingbird's 600 beats per minute), and they must eat continuously. Their diet consists of insects, worms, mollusks, and the bodies of dead creatures.

Opportunity Knocked

Primitive mammals not much different from the rodents in your backyard scuttled about at the feet of dinosaurs during the Cretaceous, 136 to 65 million years ago. The Cretaceous was a time of giants. Not only were the vast majority of land animals bigger than today's horses and cows during the reign of the dinosaurs, but forests were dominated by mature conifers that soared upward like redwoods, their canopies arching over their domain in cathedral-like majesty. Both flowers and mammals were rare. They took ecological "leftovers" and made a living. Flowers claimed disturbed, open ground or perched on mountain slopes, while most mammals scurried about in the dark, learning to be quick and clever and sharpening their sense of smell.

Then the climate slowly worsened. There were greater seasonal variations in temperature. Shallow seas began draining and mountain building decreased, which had the effect of reducing the number of different habitats available for exploitation. About 65 million years ago, it appears that an asteroid colliding with the Earth may have brought on a "miniwinter" of its own as tons of debris filled the sky for days or weeks or months and blotted out the sun.

Suddenly, a different set of animals and plants had the tools necessary for survival in a world vacated by 70 percent of its former occupants. The result was an evolutionary explosion that peaked some 27

A Mammal Family Portrait

million years ago in an age dominated by smaller, but high-energy creatures like herbaceous flowering plants, mammals, and social insects. This most recent geological period, called the Cenozoic, is broken down into two major periods: the Quaternary, which covers the last three million years; and the Tertiary, which stretches from there back to the Cretaceous. The Tertiary is in turn broken down into epochs; these are marked in the sediments by major changes in animal and plant groups that may have been initiated or augmented by other collisions with extraterrestrial debris.

For mammals there were two major trends established at the beginning of the Cenozoic that are still evident today: the development of hoofed, grazing animals and the carnivores that feed on them. Mammals also got progressively larger, culminating in giant forms like *Baluchatherium*, a hornless, giraffe-like rhinoceros that was 18 feet tall at the shoulder and 27 feet long.

Modern rodents, including the friendly mice that scuttle about near the trash cans, got their start 38 to 26 million years ago in the Oligocene Epoch. In what is now South Dakota and Nebraska, they shared

Figure 14-4: Baluchatherium, *a relative of the rhinoceros, was the largest known mammal to ever walk the earth. The head of a six foot tall man would reach a little above* Baluchatherium's *knee cap. (Based on drawing in Carrington, 1963)*

the world with the ancestors of dogs, tapirs, camels, rabbits, and hyenas as well as sabre-toothed cats, rhinoceroses, and huge titanotheres, rhino relatives with a pair of blunt horns.

In the succeeding Miocene Epoch the climate got much drier. Forests died back, and a variety of flowering plants, the grasses, took advantage of the open spaces and formed grasslands. In Nebraska, *Parahippus*, the early three-toed horse, lived alongside large wolflike dogs, rhinoceroses, tapirs, beavers, and camels. In Africa, great apes were as plentiful as monkeys are today, and among some of them began the glimmers of self-awareness that would ultimately prove to be more revolutionary than mere changes in body form.

The Mammalian Toolkit

Domestic animals provide, for most of us, our most extended opportunities for observing the behavior of mammals. They can show us quite nicely some of the things that have helped to make mammals so successful. Kids like cats because they are soft and "fluffy," and cats appreciate their "fluffiness" when Mother tosses them out into the snow for attacking the pot roast. Hair is characteristic of mammals and provides the insulation that makes their warm-bloodedness a viable option in terms of the energy expended to achieve it. Both mammalian hair and bird feathers may be highly specialized derivatives of a reptile's scales.

Most people get an opportunity to see dogs or cats give birth. Although they may not have many young compared to a fruit fly or a fish, they suckle their offspring with milk-secreting mammary glands. More importantly, they protect and teach their young over a period of months or years until the youngsters can get by on their own. Therefore, their individual survival rate is high.

If the opportunity arises, watch your female cat defend her kittens against Fido. Long before he gets too close, she will go out to meet him. She will gallop forward with combined up-and-down and sideways motions that maximize her apparent size. You can also see similar behavior when kittens play. Males will never use the behavior sequence as adults, but for females it's good practice for their motherhood. If Fido is not intimidated, the cat will unsheath her claws and go for the nose and the eyes. Few animals will stand up to the onslaught. One naturalist described an episode in Yellowstone Park of an enraged mother cat actually treeing a bear.

The long period of education that mammals go through after birth would seem to have both advantages and disadvantages. A young cat is more vulnerable than a newly born alligator, for example, because the latter is equipped with teeth and an instinctive repertoire of behavior that allows it to go out and find its own

Comparing Teeth

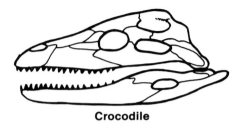

Crocodile

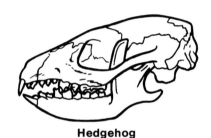

Hedgehog

lunch. The adult cat, however, is armed with learned responses to a variety of situations that make it more responsive to its particular environment. Moreover, it can respond to new dangers much more quickly. Although I've not done a survey, I suspect there are fewer puma skin rugs than alligator handbags.

The importance of learned behavior to mammals is probably directly reflected in their large brain-to-body size ratio. Vertebrate brains have gradually become larger over the millenia, adding layers that have modified the basic "wiring patterns" of primitive fish. Mammals have elaborated the outermost layer, the cortex, into a twisted, cauliflower-like structure that stores and correlates learned behavior. The furtive scrambling of rodent-sized mammals beneath the shadows of dinosaurs may have forged these complex neural pathways to compensate for lack of physical size and strength.

The acute senses of hearing and smell that mammals possess are derived from their early nocturnal existence. In addition, the modification of certain bones in the reptilian jaw to form the bones of the inner ear of mammals provided a bonus; the mammal's jaw is a single bone called the dentary, which can be moved from side to side as well as forward and back. This allows for greater efficiency in

Figure 14-5: Mammalian teeth, like those of the hedgehog, are highly adapted for chewing, with complex, shearing surfaces that must fit correctly together. Reptilian teeth are simple spikes that can be regrown when they wear out. Greater efficiency in chewing in mammals goes hand in hand with more efficient digestion, which helps to maximize the energy received from food. This is important in maintaining endothermy ("warm-bloodedness"). (Redrawn from Young, 1981)

breaking up food, which in turn makes food more easily digestible.

The teeth that line the jaws are also a considerable improvement over the peg-like teeth of reptiles. Reptiles need sharp, well-developed teeth from birth, as most must immediately fend for themselves. Since mammals are fed initially on milk, other options are open to them. Mammals shed one set of teeth after weaning, and the permanent ones that follow have complex shearing surfaces that efficiently grind seeds and nuts. Reptiles, which can continue to grow throughout their lives, replace their less complex teeth in un-limited number. I have to admit that I sometimes envy those reptiles when I look into my own mouth and see all those "complex shearing surfaces" carefully re-created in silver and gold.

Mammals really took a step forward, however, in their mode of reproduction. Reptiles pioneered the shelled egg, a structure which removed their close de-pendency on water for reproduction. Mammals took this a step further by re-taining the egg in the mother's body, in essence making the young embryo a para-site of the mother until it reaches an advanced stage of development. Two modern animals that are relics of an earlier time are the duckbilled platypus and the echidna. Both have hair and suckle their young, but lay eggs in the reptile's fashion.

The Cretaceous saw at least two and perhaps three major reproductive exper-iments among mammals. There were placental mammals, like ourselves, in which the embryo ties into the mother's blood supply via an umbilical cord that is connected with the placenta. The pla-centa is a fusion of tissues from embryo and mother where food and waste ex-change can occur. Marsupial mammals also lived during the Cretaceous. The young of these mammals migrate out of the uterus at an early stage and crawl into a teat-lined pouch where development is completed. Finally, multituberculate mammals (a name which means "many bumps") were distinguished by teeth with many bumps on them. Multituberculate mammals were herbivores with rodent-like incisor teeth and other skull features not unlike those of placental plant eaters. The size of their birth passages, however, implies they had young that were very small at birth, like marsupials. Although they survived the Cretaceous extinction, they died out more than 24 million years ago in the Eocene, perhaps because of competition with placental and marsupial mammals.

Most mammals alive today are placen-tals, although relic populations of mar-supials survive in Australia and South America. There are a few quite unique species of marsupials, like kangaroos, but most eventually filled niches similar to those of their placental cousins and now resemble placental mammals with similar "jobs." It has been common to think of marsupial mammals as "inferior" to pla-centals because placental types have tended to supplant marsupials in mixed populations. However, some biologists have pointed out that throughout history marsupial populations, isolated on the island continent of Australia and the tropical areas of South America, were subjected to less intense selection pressures than placentals. Thus, their failure to compete with placentals may reflect this history of diminished competition, and a less varied gene pool to draw upon because of their relative isolation.

Chance may have played a large part in determining which mammals entered the Cenozoic eden after the Cretaceous disaster. Only a few marsupials made the transition. Would marsupial primates have explored the road to self-awareness and, if so, would human history have been substantially different because of such a variation in how we bore our children?

Growing Up in the Ice Age

As mentioned earlier, the Cenozoic was a period of climatic decline. Thirty-seven million years ago winter seems to have been reinvented after hundreds of millions of years of moderate temperatures. This juncture, ending the Eocene and beginning the Oligocene, might also have been marked, like the end of the Cretaceous, by the fall of an asteroid. At any

rate, seasonal frosts came to northern latitudes, sea temperatures dropped, and glaciers grew in Antarctica. Amidst this climatic turmoil the first recognizable dogs and cats appeared.

People, of course, were still far in the future at this point of evolutionary history. Grasses would not even appear for another thirteen million years; they were eventually forged from the demands of the drier conditions that existed 24 million years ago. Temperatures moderated for the next ten million years, though still cool by dinosaur standards. Grass eaters, their predators, and all kinds of apes and monkeys flourished when grasslands created new habitats. Humans, like the dogs and cats before them, were products of another deterioration in climate that culminated in a series of "ice ages" beginning three and a quarter million years ago. Curiously, this period, the Pleistocene, saw a case of "giantism" among mammals, with many species of elephants, horses, bison, wolves, deer, bear, and other animals growing to giant proportions. Most of these forms died out 10,000 years ago, at the end of the last major glaciation. Perhaps the most notable aspect of this extinction is that a new force is implicated: the power of self-aware minds. You may wish to look up the work of Bjorn Kurten, who has been a long-time advocate of the human role in Pleistocene extinctions.

A Dog's Life

Dogs may not be entirely blameless for the Pleistocene extinctions, because their collaboration with mankind goes back at least 30,000 years. Dogs would have been useful in the prehistoric "round-ups" that were known to have sent entire herds of mastodons to their deaths over a convenient cliff. Most modern dogs have a predominantly jackal ancestry. Their forebears were undoubtedly camp followers that took advantage of the scraps of food left behind by bands of early humans. Perhaps humans derived some benefit from the warning cries of half-tame jackals when predators approached. Jackals may have learned, after long years of association, that a meal could be had by harrassing a large animal long enough to bring the noise to the attention of the clever apes with the lousy sense of smell. Or perhaps a human child, liking things soft and fluffy as they do, rescued a parentless pup who grew up attached to a human being instead of another wild jackal— a case of mistaken canine identity. It wouldn't take long to discover that a tame jackal that thought you were its mother could be a valuable hunting partner.

The loyalty of dogs apparently has two sources: a youthful retention of the bond between mother and pup into adulthood

Figure 14-6: *Sheena, a rock-retrieving friend of mankind.*

and the pack loyalty of a dog for its leader. The latter characteristic is especially true of dogs with some wolf blood in them. There is quite a range of personalities that arise just from the breeds with jackal ancestry. Ultra-friendly dogs that greet every human being as a long-lost relative have a large dose of the juvenile-mother bonding instinct. These dogs are easy to get along with but are pretty fickle. We have one dog, a black lab, that leans heavily in this direction. We acquired her as an old dog from people leaving the state who couldn't take the dog with them. The dog, Sheena, had grown up with the children of that family, and yet she adapted almost immediately to a new set of adults and children. Her only requirement in life is that you feed her and give her rocks to chase and fetch on a regular basis. Our other dog, Trash, is part black lab and yet has a totally different personality. We raised her from a pup. She is less cosmopolitan with her affection and tends to treat us as somewhat more powerful equals. Her name, by the way, refers to her willingness to eat everything, rather than a commentary on her general worth.

Some dog breeds have a bit of wolf intermixture. For them, loyalty springs from an allegiance conferred on a suitable human in lieu of a pack leader. The period for this bonding may be very short. Between the fifth and seventh months a pup will select a pack leader if a suitable "model" is present. The bonding itself can occur within a few days and last a lifetime. Konrad Lorenz, a famous student of animal behavior, describes many of his experiences with dogs in a delightful book, *Man Meets Dog*. One dog, an Alsatian-Chow mixture whose name was Stasi, bonded to Konrad at the age of seven months. He trained her over a two month period. She was completely housebroken, ignored the poultry, and obeyed a variety of verbal commands. Then, because of a job offer that separated Konrad from his family for nine months, the dog went virtually wild, killing livestock and terrorizing the neighborhood. Yet when Konrad returned, the dog's "mental stability" returned immediately. Stasi immediately obeyed all the conventions and commands she had formerly learned.

Such dogs that don't bond to an individual may become very independent and emotionally insulated from humans. They may still live comfortably with people, but treat them as co-equals and acquaintances rather than masters. It would be interesting to know if wolf pups isolated from other wolves after birth would show some of the same social deficiencies as primates so deprived. The now classic experiments of Harry F. Harlow showed that monkeys must have the opportunity to interact with their mothers and other siblings in order to be able to form successful relationships later in life. If deprived of this contact for the first six months of their lives, they become permanently unable to function normally with other monkeys. Harlow also looked at institutionalized humans and found that the effects of isolation could be reversed for them if normal social contacts resumed within the first six months of life, but emotional scarring got progressively worse after that until, by two years of age, many children had reached "the point of no return."

Perhaps the strongest common tie between man and dog is our mutual need for social interaction. We both need love and have to learn how to give it early or we lose the capacity.

And Then There Are Cats

Cats are something else again. Lion prides notwithstanding, cats tend to be far less social animals than dogs, and I think our affection for them has different roots. To a certain extent our feelings for cats may be based on inaccurate perceptions of what's going on in their heads. Cats appear "proud," "independent," "aloof," and "cunning." By inference from these apparent traits we project a certain intelligence on them which, alas, they don't have.

Cats give straightforward signals of their feelings. As they get more perturbed at something you're doing, their ears become more laid back, their eyes become less round and more almond-shaped, and the tip of the tail flicks back and forth like a metronome. The striped patterns of the basic gray tabby enhances facial signals and makes them unmistakable. If you persist in your attentions and the cat can't go elsewhere, it will lower its body and emit a low growl. Ultimately it will lay on its side, flicking its tail and presenting all its weapons in an "I dare you" posture.

Cats will also defy you to your face. I have one cat which clearly displays her intention to jump on the table or counter while I watch, a forbidden act in my house. She does have the good sense to scram, however, when I make the slightest protest. My dog, Trash, is similarly forbidden to enter the garden. She wouldn't think of crossing my wishes in open view, unless under the duress of spotting a foreign cat there, but she will wait until I'm inside and, after a glance at the window to make sure I'm not watching, enter the "Nodogsland." A dog's lies demonstrate their intelligence as much as their happy-go-lucky "Yes, sir, anything-you-want, sir" exuberance seems to belie it.

We owe our relationship with cats, of course, to the Egyptians. Cats seemed to represent the majesty of lions to the Egyptians, and they became sacred to the goddess Basd. Killing a cat in the 5th or 6th dynasties carried the death penalty. Their practical worth in killing rodents undoubtedly aided the cat's popularity. Cats were as carefully embalmed and entombed as their human companions. Weasels were used to control rodents in Europe, although they were much harder to domesticate. Plutarch says cats were

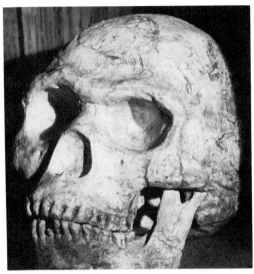

Figure 14-7: *Skhūl V belongs to a population of humans very similar to modern man (and is considered a member of our species), but several skull measurements are outside the range of modern populations. Skhūl V might be considered an example of a man "undomesticated" by civilization.*

Figure 14-8: *This clay reconstruction of Skhūl V was created by the author under the direction of the physical anthropologist, Dr. Michael Charney, who has done many facial reconstructions of murder victims at his laboratory on the campus of Colorado State University.*

introduced in Europe in the first century A.D., and pet weasels couldn't quite compare after that.

By the 12th and 13th dynasties, as demonstrated by embalmed specimens, cats were showing signs of domestication: ear structure and coloration changed, the temporal bones of the skull became more domed, and their noses were stumpier. Similar changes in dogs 30,000 years ago were displayed by fossil "turf dogs" found in Europe in association with human communities that lived in stilted houses and fished the sea.

Self-Domestication

One of the human skulls in my collection is designated as Skhūl V. His remains were those of the fifth individual found in a cave burial site unearthed in Israel on the western slope of Mount Carmel between 1929 and 1934. Skhūl V lived between 35,000 and 45,000 years ago. There were no grave goods with the two women, five men, and three children found there, but they were laid out in order and the jawbone of a wild boar was found in the arms of Skhūl V. Skhūl V was basically modern in form, although he shared some traits with the classic "Neanderthal man" that lived in a western Europe dominated by the icy breath of glaciers. When I compare Skhūl V with a modern skull, I am struck both by the similarities that bind us as well as by the differences that, to a large extent, are similar to the differences between wild and domestic animals. We, as a species, have been domesticated by our own construct: civilization.

Skhūl V, for example, has a more massive and robust skull than a modern man. The widest part of his face is 5/8 inch broader than my modern sample. His jaw is less delicate. The ramus of the jaw, which is the portion that leads up to the hinge with the top of the skull, is thicker.

Larger teeth are well worn, but with no cavities. Skhūl V did suffer from pyorrhea and a couple of abscesses, however. The arch of bone that connects the outer part of the eye sockets with the area near the ear opening is heavier. At the rear of the skull, a bony ridge called the nuchal crest, where neck muscles attach, is more pronounced, indicating that the muscles were well developed. Most noticeable to the casual observer, the brow ridges are relatively heavy and would have shadowed the eyes of Skhūl V.

Domesticated animals show various hormonal changes that involve temperament and reproduction. Egyptian cats, for example, were much easier to live with than European wild cats, which was one of the reasons they were the "mouser of choice" in Egypt, whereas in Europe the weasel guarded the pantry. In successive generations, docility was further enhanced by inbreeding only the mildest personalities. Artificial selection for certain traits unmasks various hidden or recessive genes and results in a much more rapid evolution than natural selection, which is limited to picking survivors in a slowly shifting succession of environments. Somewhat more surprising is a shift in hormones important in reproduction. Whereas wild animals are usually locked to the seasons with regards

Skhul V and Modern Man

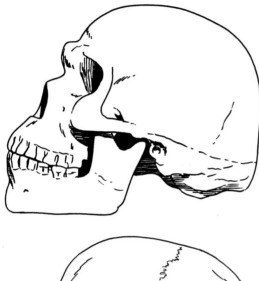

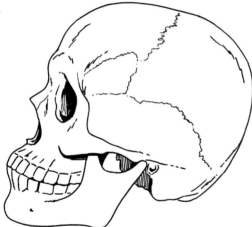

Figure 14-9: While looking at the skulls of Skhūl V (top) and a modern man (bottom), compare the following: the general flatness of the face, the amount of skull "doming" above the brows, the heaviness of various features like brow ridges, jaw and sites of muscle attachment, and the size and placement of teeth in the jaw. Domestic animals show somewhat analogous changes from their wild counterparts.

to mating and childbirth, domestic counterparts can breed all year long. Continuous breeding is also a characteristic of human populations, and it raises the question of how much of our behavior can be attributed to original biology and how much is an artifact of our control over our environment.

Artificial selection has been very efficient at achieving short-term ends. The wild auroch, ancestral to the cow, may have given only a few hundred grams of milk, whereas the best milkers today can give 12,000 to 15,000 liters per year. Wheat and corn have been bred for large yields and for seeds that stay bound to the stalk for easy harvesting. However, these various animals and plants are now bound permanently to human intervention because they can't survive without us. Whereas survival was the criterion of natural selection, human utility is the criterion for artificial selection, and the two are not always compatible. Domestication was a great achievement for human beings, but it's important to remember that our choices for further domestication are restricted by the pool of wild animals and plants from which we have to draw.

Skhūl V lived among wild ox, hyena, hippopotamus, rhinoceros, wild ass, gazelle, fallow deer, roe deer, red deer, boar, and small wild cats. For the most part, I share the world with dogs, cats, and mice and go to special parks to see the rare wild creatures that we, as a species, have shoved into ecological corners. Obviously, we can't go backward. Skhūl V lived a short, hard life. Agriculture and animal and plant domestication gave us the wealth and the time to build civilizations, and civilizations give individuals the time to create as well as merely survive. However, we do have to realize that variety among animals and plants is what creates the raw material for survival, and if the circle of our natural allies becomes too small, our tenure on Earth will not be nearly as successful as that of our dinosaur relatives.

REFERENCES

Burt, William H. 1967. *Mammals of the Great Lakes Region.* Ann Arbor, Michigan: University of Michigan Press.

Carrington, Richard. 1963. *The Mammals.* New York: Time Inc.

Day, Michael H. 1977. *Guide to Fossil Man,* 3rd ed. Chicago: The University of Chicago Press. A comprehensive listing of human fossil remains and a description of the findings.

Durrell, Gerald. 1983. *A Practical Guide for the Amateur Naturalist.* New York: Alfred A. Knopf.

Harlow, Harry F. 1966. The Nature of Love. In *Human Development,* 2nd ed., Haimowitz and Haimowitz, eds. New York: Thomas Y. Crowell Co.

Harlow, Harry F. and Margaret K. 1966. Social Deprivation in Monkeys. In *Human Development,* 2nd ed., Haimowitz and Haimowitz, eds. New York: Thomas Y. Crowell Co.

Lewin, Roger. 1982. *Thread of Life: The Smithsonian Looks at Evolution.* Washington, D.C.: Smithsonian Books.

Lorenz, Konrad. 1953. *Man Meets Dog.* Baltimore, Maryland: Penguin Books Inc. A delightful, short book that anyone interested in dogs and cats should read.

Palmer, Ralph S. 1954. *The Mammal Guide.* Garden City, New York: Doubleday and Co.

Stanley, Stephen M. 1986. *Earth and Life Through Time.* New York: W.H. Freeman and Co. A new textbook for the earth sciences that does a good job relating biological evolution to geological processes and events.

Whitfield, Philip, ed. 1984 *Macmillan Illustrated Animal Encyclopedia.* New York: Macmillan Publishing Company.

Young, J.Z. 1981. *The Life of Vertebrates,* 3rd ed. Oxford: Clarendon Press.

15

Man Sorts Out Nature

Nearly everyone, at one time or another, is awed by the incredible diversity and abundance of the living world. Various people, at different times in human history, have attempted to make some sense of this overwhelming bounty. All attempts have fallen short of total satisfaction, but each try has resulted in more information about the kinds of creatures with which we share the world. This final chapter describes important changes in man's view of his place in nature and serves to illustrate the thinking that led to the current system (see Chapter 2) used to classify the many plants and animals you have observed and studied.

The Bookkeeping Problem: What Goes Where?

The urge to organize seems to be a characteristic of the human condition. Undoubtedly, plants and animals were some of the first things to be "classified" into different groups. A simple form of classification might result in such categories as "food items," "medicines," "poisons," and "hallucinogens." As in all systems, boundaries are not firm. Food items could also be poisons or hallucinogens under the right conditions. Many classification systems are possible. We could arrange organisms by their uses, by their colors, alphabetically, by their size or whatever. Useful classification systems, however, should accomplish several things: you should be able to place newly found organisms within them easily, organisms should not fit in several different places, and members of a group should have similarities that are basic rather than superficial (such as size, color, or alphabetical groupings).

Structural similarities and differences are apparent among organisms. Moreover, if you collect large samples of animals and plants, you can begin to see a continuum of forms ranging from simple to complex. At the "simple" end, for example, traditional boundries between plants and animals become fuzzy, whereas at the other end you have no trouble telling a petunia from a porcupine. The arrangement of creatures by structural similarities and level of complexity led to the concept of "The Great Chain of Being," which took several forms. Sir Thomas Browne expressed the concept well in his *Religio Medici* of 1642: "There is in this Universe a Stair, or manifest Scale of creatures, rising not disorderly, but with a comely method and proportion."

The Great Chain of Being was a static concept in that it postulated an eternal order of things. Life was forever bound in a hierarchy of complexity with man at the top. "From Nature's chain whatever link you strike," Alexander Pope said, "tenth or ten thousandth, breaks the chain alike." It was often a strain to fit an organism within such a static chain. Is an ant, for example, more or less complex than a frog, or is it just different?

Nevertheless, arranging organisms by their structure seemed a valid way to make sense of things, and Carolus Linnaeus in his *Systema Naturae* of 1758 set up a two-name system that we still use today. In his scheme of things each organism got a species name and a genus name. The species "unit" was the smallest grouping in his system and included individuals that were all of the same kind. Members of a species only breed with each other. The genus group represented clusters of species that were similar, yet distinct. Thus man was given the name *Homo sapiens*, or "Man, wise." Linnaeus, apparently on the basis of travelers' descriptions of orangutans, also listed a *Homo troglodytes*, or "Man, cave dweller."

Linnaeus' broadest classification grouping, except for the kingdoms of plants and animals, was the class. Classes included smaller groupings called orders, which in turn contained families of genera and their individual species. Later, larger units than the class were introduced: the phyla. Each organism, then, was placed in a kingdom, phylum, class, order, family, genus, and species. A pnemonic device for remembering the sequence might be "Kings play cards on fast, gray ships". Thus, our own personal spot in the classification system becomes: Kingdom: Animalia; Phylum: Chordata (animals with backbones); Class: Mammalia (hairy, live-bearing, young are sucklers); Order: Primates; Family: Hominidae (a family we have to ourselves at present); Genus: *Homo*; Species: *sapiens*.

The Problem With Fossils

The Great Chain of Being, like most attempts to simplify nature, eventually ran straight into reality. One thing it couldn't explain was fossils, and a few people even tried to explain them away as artifacts placed on the Earth by the Devil to confuse matters. Many fossils represented animals and plants that still existed or existed in only slightly altered form, but others were the remains of strange creatures that seemed totally unique. Eventually it became clear that whole ecologies once existed that were undreamed of. How could they fit into the Chain of Being?

Georges Cuvier (1769–1832), the premier biologist of his day, addressed the problem of fossils. His insights came from the comparative study of animals. In fact, he is considered the founder of comparative anatomy. His work led to the expansion of Linnaeus' classification system to include phyla. He divided the animal kingdom, for example into four phyla: Vertebrata, Molluska, Articulata, and Radiata. Cuvier recognized that fossil animals shared structural similarities with living forms and could be placed in his classification system.

It was not difficult to relate fossil elephants or ground sloths to their living relatives. Most fossils that are found in the uppermost layers of rock fall into this category. The creatures they represent are minor distortions of modern forms. As you dig deeper, however, the animals and plants show greater variations. Nevertheless, when the first pterodactyl was discovered in the early 1800's Cuvier could point to structural characteristics and identify it as a flying reptile.

Cuvier realized that the many layers or strata of earth, each containing its own distinctive array of fossils, must represent a vast amount of time. He desired a way to explain things that would not interfere with his devout religious beliefs, which included the biblical record of Genesis. What he decided was that the Earth had suffered a series of catastrophes, the last of which was the biblical account of the Flood. The force of his achievements,

reputation, and personality made the catastrophism theory popular until his death.

Sir Charles Lyell (1797–1875) met and was impressed with Cuvier, but he would eventually popularize a concept that would overthrow catastrophism. Lyell was a geologist who traveled widely in Europe, visiting various geological features. He became convinced that all the physical features on the Earth could be explained by the action of physical forces: wind, water, earthquakes, and vulcanism. The only additional ingredient needed was sufficient time. By measuring the rate of action of these forces you could determine the amount of time needed. Lyell estimated the age of the earliest fossil-bearing rocks to be 240 million years, which was a vastly larger estimate than any other at the time, although it is less than half of current estimates.

Modern dating techniques rely on several methods, including counting tree rings, measuring the rate of radioactive decay of certain elements and isotopes, and noting the location of fossils relative to rocks that give evidence of changes in the Earth's magnetic field. All the techniques support the notion that the Earth has been around some 4 billion years and that life has existed for at least 3.5 of those billions.

Lyell later found that his ideas, called

uniformitarianism, were initially proposed by a geologist a generation earlier, James Hutton. Lyell's book, *The Principles of Geology*, was so well written and popular, however, that it came to have a great impact on other scientists, including a young Charles Darwin.

How Do Species Change?

Charles Darwin (1809–1882) was the man who made the concept of evolution viable. He was not the first to consider the idea. The great French naturalist of the previous century, Georges Louis Leclerc de Buffon, is one example of a person who proposed evolutionary explanations for the differences in animal and plant species. ". . . If the point were once gained that among animals and vegetables there had been . . . even a single species which had been produced in the course of direct descent from another species . . . then there is no further limit to be set to the power of Nature, and we should not be wrong in supposing that with sufficient time she could have evolved all other organic forms from one primordial type." Buffon, however, was making speculations from informal observations, and the concept was not at all officially popular.

One thing that Darwin accomplished was the marshalling of so many observations and examples of evolutionary change that the transformation of one species to another became the best explanation for what was occurring and had occurred in nature. During his five years as a naturalist aboard the *H.M.S. Beagle* he collected and observed thousands of plant and animal species. Many of the observations that became central to his theory took place on the Galapagos Islands. "Darwin's finches" are legendary examples of how one "founder" population of animals can change to fill all the available "jobs" or niches in an environment where there is little or no competition. Darwin drew heavily on parallels between what man accomplishes in plant and animal breeding and what happens in nature. Random variations produce varieties that people can enhance by selective breeding practices. Darwin pointed out, too, that many organisms contain vestigial organs or other parts that serve no purpose but may reflect structures that once were important to survival.

Darwin also proposed a mechanism for this process of transformation: natural selection. He did not arrive at this theory immediately after his voyage, however. Perhaps the precipitating event was when he read *An Essay on the Principle of Population* by Thomas Malthus. Malthus proposed that all populations increase to the limits of their food supply. After that, disease and starvation, or even war in the case of humans, trims down the surplus. Obviously those best suited to cope with these stresses would survive more often than those who weren't. Thus, Darwin came to believe that nature selects the fittest organisms from overly large populations to survive and pass on their characteristics to their offspring.

Exactly how organisms passed on such traits was a weakness in his theory. The notion of inheritance current at the time was that traits from both parents blended together in their offspring. This would tend to dilute any favorable characteristics from one parent. It wasn't until the work of Gregor Mendel was rediscovered that the basically particulate nature of inheritance was recognized. Traits could be passed on intact from one generation to another, sometimes completely masked in one generation only to appear in the next.

Darwin also considered the possible role of sexual selection. Females may select males on the basis of coloration, behavior, or some other trait that may or may not reflect fitness in a particular environment. Bright coloration, for example, may appeal to a predator as well as a mate.

The essence of evolution by natural selection as proposed by Darwin, however, was that it was a slow process, fueled by random variations that occurred in populations. Modern versions also emphasize the importance of isolation. Isolation can be caused by physical barriers such as mountains or water, or it can be caused by behavioral traits that discourage matings between groups. Isolation provides the time for small variations to accumulate until the differences between groups are so great that we recognize them as different species.

Today there is some argument about the speed of evolution. Is it always slow or can it sometimes be quite rapid? Stephen Jay Gould of Harvard University is a proponent of a "punctuated equilibrium" theory. The idea here is that a relatively major change may occur once in a great while, perhaps by a mutation during an early stage in embryonic development or in a gene sequence that controls a constellation of hormones or enzymes, and that isolation and speciation can occur in several or even one generation. Between these events, evolution proceeds more leisurely, carving out individual species through less drastic mutations.

A New Catastrophism

Strangely enough, the importance of catastrophes is again taking on importance in modern discussions of evolution and the history of life on Earth. Much of this began several years ago when Luis Alvarez of the University of California at Berkeley began studying the chemical composition of the layer of clay that separates two major periods in Earth's history: the Cretaceous and the Tertiary. This boundary has long been of interest because 70 to 75 percent of the species living during the Creataceous died at the end of that period, including the dinosaurs.

What Alvarez found in this clay boundary was a very high level of iridium, a metal rare on Earth, but not uncommon in extra-terrestrial bodies like asteroids. The initial studies were done in Italy, but the same high levels of this metal were found at Cretaceous-Tertiary boundaries world-wide. Alvarez speculated that an asteroid may have collided with the Earth, throwing huge quantities of dirt into the strato-sphere, significantly blocking sunlight for an extended period of time—perhaps even months or years. The result was that land plants and ocean plankton died first, with animals higher in the food chain following shortly thereafter.

In the fall of 1983 paleontologists David Raup and John Sepkoski of the University of Chicago announced the results of a study of mass extinctions over the last 250 million years, a span of time marked by the extinction of some 3,500 families of marine organisms. To their surprise a distinct pattern emerged. Mass extinctions peaked sharply at intervals of 26 to 28 million years. Periodic disasters appear to significantly affect the course of evolution on Earth. Although no one has discovered the cause of this cycle of disasters, one theory is that our sun may have a com-panion star (double stars are the rule rather than the exception in the galaxy). This companion star could periodically come close enough to the cloud of rocks and ice believed to exist on the outskirts of the solar system to nudge some of this material in toward the planets closer to the sun. The result would be a periodic increase in debris near the Earth. Statis-tically that would translate into more hits by meteorites, some of them asteroid-sized.

What this all implies is that the Earth is "cleaned" of much of its life from time to time, and the vacancies that are left in various habitats are filled by the survivors, much as a few finches found the Galapagos Islands and diversified into fourteen different species. Mammals, in-cluding ourselves, may owe their ex-istence to a chance event that scoured the very successful dinosaurs and their kin from the face of the Earth.

The Role of Intelligence

Whether or not human beings are around today because of an ancient "falling star," chances are that our capacity for language, self-awareness and manipulation of the environment would have shown up eventually in some creature at some time. Intelligence (which I'm roughly equating with these three capacities) has shown itself to have survival value—at least over the "short-term" lifetime of our genus, which may be several million years. The dinosaurs, in fact, might have explored our ecological niche if they had been given the time. At the end of the Cretaceous there were small, hunting dinosaurs with binocular vision, relatively large brains, and forelimbs quite adept at manipulation. One of these Coelurosaurs could have been stressed by a rapidly shifting climate, as our ancestors apparently were during the Pleistocene, to marry intelligence to the use of these natural tools.

Chance will play a lesser role in determining the survival value of the human brand of intelligence over the long haul of geologic time. Certainly our manipulation of the environment and the elemental forces that mold the form of the universe can help us counteract some of the "cosmic" disasters that could befall us. Even with today's technology we could divert an asteroid from a collision course—once we saw it coming. However, the odds for that sort of disaster happening anytime soon are exceedingly small. The greatest dangers are those we create for ourselves through ignorance, apathy and the pursuit of certain misguided instincts. Intelligence has short-circuited the leisurely pace of evolution by substituting learning and the accumulation of knowledge for physical and behavioral changes wrought by changes in our genetic make-up. This substitution could be a very efficient one—*if* we can learn quickly enough which "instincts" to suppress or reroute.

Wars, for example, in the form of border skirmishes between tribes, may have been relatively harmless ways to defuse aggressive instincts that could have been turned against family and friend. They may have been mechanisms for selecting the strongest or cleverest individuals to seed the next generation. They may have served to space out human communities so that environments weren't taxed beyond their limits. But in today's world, with today's weapons, wars can only contribute to fouling the entire planet and hastening our extinction. We can't eliminate aggression which, when properly controlled, contributes to our ability to struggle against adversity, but we should be able to discover ways to channel it properly by using rather than abusing our intellectual powers.

All animals have a biological imperative to reproduce. For most species there is no danger of their reproduction getting out of hand. Weak or ill-adapted offspring fall prey to accident, disease, and predation. Others migrate to virgin territory. But humans have no serious animal competitors and we threaten to fill the Earth with ourselves and our waste. Graphs of human population growth and the increase in extinctions of mammals and birds over the last three hundred years show strikingly parallel curves. We don't yet have the option of colonizing another planet and, even if we did, it would be a sad testimonial for sapient life if we couldn't manage our numbers to coexist with the one world most perfectly suited to us. Strong moral reasons, rooted in instincts for cooperation that have served us well over the millenia, make murder a reprehensible option for population control. Yet, like rats in overcrowded conditions, our hormone systems could become more and more unbalanced until violent aggression predominates. The immune system also fails such stressed animals, and disease cuts their numbers. Because deaths, births and migrations are the sole

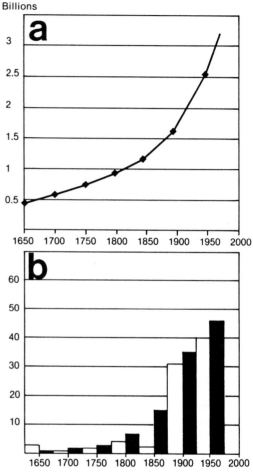

Figure 15-1: *Comparison of growth curve for human population (a) with the number of extinctions (b) of mammals (white bars) and birds (black bars) over the last three hundred years. Redrawn from Ziswiler, 1967*

determinants of a population's size, the only humane way to control our numbers is by managing the number of births. Hopefully, intelligence will give us the will to achieve that goal before more ancient solutions prevail.

Most creatures need not consider the problems of generating too much waste. The natural world is well suited to re-cycling wastes at a tempo commensurate with the speed at which they are pro-duced. Yet over the long haul, this may not be the case. When primitive microbes used up the "organic soup" of the ancient oceans, a crisis was generated that led to the use of light energy to produce neces-sary carbon compounds. Tapping this energy source, however, produced the pollutant, oxygen, that was poisonous for most life of that era. Humans, too, are tapping many energy sources and pro-ducing vast quantities of waste: carbon dioxide, non-degradable plastics, sulfur dioxide, radioactivity and ozone-eating chlorofluorohydrocarbons, to name a few. The difference, in our case, is that we are generating these pollutants at a rate im-possible for nature to cope with. If we do not solve these pollution problems, the Earth of future eras might be restabilizing itself after the fall of human civilization.

Most organisms, too, are naturally self-ish. They thrive at the expense of those neighbors they compete with. No species before man, however, has been so domi-nant that it threatened to fill all the various habitats of our planet. Mankind must

unlearn that selfishness. Nature is not an adversary to be conquered, but another manifestation of ourselves that is vital to our well-being. Edward O. Wilson believes we all share a natural attraction to living things which he terms biophilia: "To an extent still undervalued in philosophy and religion, our existence depends on this propensity, our spirit is woven from it, hope rises on its currents." Our failure to fully understand the intricate intercon-nections of the living world blinds us to the harm we're causing as the most potent exterminators of all time. More pervasive than a changing climate, faster than a speeding asteroid, Super Man is "simplify-ing" the world to death by the gradual elimination of all species but his own. Most of the destruction we humans are responsible for is not intentional. Cer-tainly, we have consciously eradicated close animal competitors and pests, but in most cases we destroy whole ecologies by destroying habitat.

The clear-cutting of tropical rain forests is the most tragic example of this in our time. In forests that house approximately fifty different species of organism per tree, the World Wildlife fund estimates we are losing twenty-five to fifty acres of forest a *minute*. Adding to the tragedy is the fact that these lands are virtually irreplaceable. Unlike temperate forests with rich layers of soil and humus that can nurture fallen seeds, tropical forests have most of their organic resources tied up in the trees themselves. The soil that is left after cutting is thin, mostly clay, and is easily washed and blown away. The loss of species is staggering, especially when we consider that two thirds of the animal life in tropical forests exists in the forest canopy and is virtually unknown. What we lose, in addition to the beauty of a varied world, is a rich pool of potential resources. In the fifty or so acres of forest destroyed during the next minute, will we lose a cure for cancer or a clue to the process of aging? Will we lose a plant like the copaiba tree whose sap is similar enough to diesel fuel to run a car? We owe it to ourselves and future generations to save enough forest enclaves to ensure a diversity of forest organisms. Just as the sound of one note is not music, a planet with only one species is not alive.

This doesn't mean, of course, that every species can or should survive. Extinction, adaptation and change have always been the rule rather than the exception, but we should not assume that every creature is expendable until proven "useful." If we have to err, it should be on the side of conservation. Just as we don't need a college degree in nutritional biochemistry to eat a well-balanced diet, we don't need

arcane knowledge in biology to accept and encourage diversity in the life around us. We all draw our emotional well being, even if unrecognized, from the life around us, says Edward Wilson, and "to the degree that we come to understand other organisms, we will place a greater value on them and on ourselves." If enough people discover the enjoyment, beauty and knowledge offered by the diversity of life in their own backyards, then we as a species may be even more successful than the dinosaurs in our gardens.

REFERENCES

Asimov, Isaac. 1972. *Isaac Asimov's Biographical Encyclopedia of Science and Technology.* New York: Avon Books. Asimov discusses all the scientists involved in the Great Chain of Being concept, but was particularly useful with Linnaeus, Cuvier, Lyell, and Darwin.

Durant, Will and Ariel. 1965. *The Age of Voltaire.* New York: Simon and Schuster. Interesting material on Buffon from an historian's viewpoint; the quote from Buffon is from that source.

Gould, Stephen J. 1983. Bound by the Great Chain. *Natural History*, vol. 92, no. 11 (November). This and the following article provide a good historical background on the Great Chain of Being concept. The quotes by Sir Thomas Browne and Alexander Pope are from the Gould articles.

Gould, Stephen J. 1983. Chimp on the Chain. *Natural History,* vol. 92, no. 12 (December).

White, Peter T. 1983. Nature's Dwindling Treasures: Rain Forests. *National Geographic,* vol. 163, no. 1 (January).

Wilson, Edward O. 1984. *Biophilia.* Cambridge, Massachusetts: Harvard University Press.

Ziswiler, Vinzenz. 1967. *Extinct and Vanishing Animals.* London, England: The English Universities Press, Ltd.

Appendix I

A Review of Sterile Technique For Handling Microorganisms

Bacteria and fungi can be grown on plates of agar, a powdered alga derivative that gels after being dissolved and heated in water. Often various nutrients are added to the agar to promote growth. A common nutrient agar consists of 3 g of beef extract, 5 g of peptone, and 15 g of agar dissolved and brought to a boil in 1,000 ml of distilled water.

One must always use sterile technique in handling microorganisms to avoid contamination of the cultures and the environment. You never can be sure whether or not you are handling a pathogenic organism, especially when you are sampling at random. The small, flat petri dishes used in culturing can be purchased from supply houses as "one shot" sterile plastic dishes or reusable glass ones. Glass dishes can be sterilized by wrapping in aluminum foil and placing in the oven at 250°F for 30 minutes. You will also need an inoculating loop and a bunsen burner or alcohol lamp. These can be purchased from supply houses.

To begin working with microorganisms clean the surface of your work area with a disinfectant solution. Sterile technique in preparing bacterial smears on glass slides consists of the following steps:

1. First place the inoculating loop above the topmost part of the flame to prevent cells from sputtering off the loop.

2. Bring the loop to just above the inside cone of flame and heat until the loop is red hot.

3. Place two drops of water on a clean glass slide and flame the loop again.

4. Remove the cotton plug of the test tube containing the bacteria with your loop hand and flame the neck of the tube, which is in the other hand.

5. Remove a loopful of culture after cooling the loop momentarily on a vacant piece of agar. Don't dig into the surface of the agar.

6. Flame the tube again.

7. Return cotton plug to the test tube.

8. Spread bacterial cells in the water drops to the size of a dime. Allow the smear to air dry. Flame the loop, but avoid spattering.

9. Heat-fix cells by quickly passing the slide over the flame twice.

Appendix II

The Gram Staining Technique

Bacteria respond in one of two ways to the Gram staining procedure. They either retain the purple color of crystal violet and are called Gram positive or the crystal violet is washed away by the alcohol treatment and they are counterstained pink with safranin. In the latter case they are called Gram negative. A fundamental difference in cell wall structure is involved in the reaction, but acidity and age of cultures also plays a part.

The response to Gram staining is often a criterion in identification. Follow these steps:

1. Using sterile technique, prepare your bacterial smears.

2. Air dry and fix in the flame.

3. Flood the smears with crystal violet stain (available from a supply house) for one to two minutes. Do not allow to dry.

4. Pour off excess stain and wash with tap water.

5. Flood the smears with Gram's iodine solution (available from supply houses, druggists, or medical labs). Allow to react for one minute. Pour off the iodine and repeat the treatment. Again, allow to sit for one minute.

6. Wash off the iodine with water and *blot* the slides dry.

7. Hold slide at an angle and apply Gram's alcohol drop by drop until the violet stain no longer appears in the washes (10 to 30 seconds). Don't overdo the decolorization.

8. Quickly rinse off the alcohol with tap water.

9. Flood the slide with safranin (the counterstain) for 30 seconds to one minute.

10. Wash gently with water, drain the excess, *blot* dry and identify the slides.[1]

---

[1]from Koby T. Crabtree, *Fundamental Experiments in Microbiology,* Philadelphia: W.B. Saunders Co., 1974.

Appendix III

Culture Media for Lichen Symbionts

The Algal Symbiont

Bold's Mineral Medium:

Part I—Six stock solutions should be prepared. Each solution will have a volume of 400 ml and will contain one of the following salts in the quantities listed:

$NaNO_3$	10.0 g	K_2HPO_4	3.0 g
$CaCl_2$	1.0g	KH_2PO_4	7.0 g
$MgSO_4$-$7H_2O$	3.0g	NaCl	1.0 g

10 ml of each stock solution should be added to 940 ml of distilled water.

Part II—Four stock, trace-element solutions are prepared as follows[1]:

---

[1]Add 1.0 ml of each stock solution in Part II to Part I. If solid medium is desired add 15 g of agar.

1. H_3BO_3 _____ 11.42 g/l
2. $FeSO_4$-$7H_2O$ _____ 4.98g/l
   $ZnSO_4$-$7H_2O$ _____ 8.82g/l
   $MnCl_2$-$4H_2O$ _____ 1.44g/l
3. MoO_3 _____ .71 g/l
   $CuSO_4$-$5H_2O$ _____ 1.57g/l
   $Co(NO_3)_2$-$6H_2O$ _____ .49g/l
4. EDTA _____ 50.0 g/l
   KOH _____ 31.0 g/l

Trebouxia organic nutrient medium I:

Bold's mineral solution	970 ml
Proteose peptone	10 g
Glucose	20 g

Trebouxia organic nutrient medium II:

Bold's mineral solution	980 ml
Casamino acids (vitamin free)	10 g
Glucose	10 g

The Fungal Symbiont

Malt-Yeast Extract Medium:

Malt extract	20 g
Yeast extract	2 g
Agar	20 g
Distilled water	1000 ml

For a clear liquid medium substitute 20 g of malt extract broth for the malt extract and add 2 g of yeast extract per liter of distilled water.

Soil Extract medium:

Bold's mineral solution	960 ml
Soil water[2]	40 ml
Agar	15 g

---

[2]Equal parts of garden soil and water are autoclaved one hour. The mixture is allowed to cool and the liquid is then filtered until clear, autoclaved, and kept as a stock solution.

Sabouraud Dextrose Agar:

Neopeptone	10 g
Dextrose	40 g
Agar	15 g

To rehydrate the medium, suspend 65 grams of Sabouraud Dextrose Agar in 1000 ml of cold distilled water and heat to boiling to dissolve the medium completely. Distribute in tubes or flasks and sterilize in the autoclave for 15 minutes at 15 pounds pressure (121° C). The final reaction of the medium will be pH 5.6.

(From Raham, R. Gary, "Exploiting the Lichen Liaison," *The American Biology Teacher*, Vol. 40, No. 8, 1978).

Appendix IV

Techniques for Handling And Preserving Invertebrates

Protozoans and other microscopically small invertebrates may need to be slowed down for viewing. Methyl cellulose is highly viscous in solution and slows these animals down mechanically. A 1% nickel sulfate solution inactivates cilia and flagella. A 0.1% solution of KCl interferes with ciliary and muscular movements.

All these reagents may distort organisms to some extent and should be considered an aid for seeing detail rather than the only way to observe fast-moving creatures. Often just allowing the water under the coverslip to evaporate a little will narrow the amount of swimming space available and slow the animal down.

When using narcotizing agents, you may need to experiment with different solutions, depending on the organisms involved. Narcotizing agents include:

Carbon dioxide: Charged water added to culture fluid is effective for many animals, including annelids.

Ether or chloroform: Prepare a moist chamber, with a pad of cotton charged with ether or chloroform. The vapor dissolves in the water containing the specimen and anesthetizes it.

Menthol crystals: A few small crystals dropped on the water containing the organisms often prove effective.

1% potassium iodide: Add to water containing specimens. Some protozoans and rotifers respond well to it.

Chloral hydrate: Solutions of from 2 to 10% are used. Add drop by drop until the organism is quieted.

1% Neosynephrin: Add drop by drop to small amounts of water containing the organism.

Methanol: Dilute to 5 to 10% and add to liquid medium drop by drop.

Magnesium sulfate: Add a saturated solution to the fluid containing the specimen. Works well with annelids.

Ether and chloroform are about as effective as any other reagent for larger organisms. If other reagents are used, it is well to remember that you may get a distorted specimen if narcotization is carried out too rapidly.[1]

---

[1](Information based on Appendix III in Paul A. Meglitsch, *Invertebrate Zoology*, Oxford University Press, 1967.)

Appendix V

Metric-English Equivalents

Equivalents of Metric and U.S. Systems

Length

1 millimeter (mm) = 0.03937 inch (in.)
1 centimeter (cm) = 0.3937 inch
1 meter (m) = 39.37 inches = 3.2808 feet (ft)
1 kilometer (km) = 0.6214 mile
1 inch = 2.54 centimeters
1 foot = 30.48 centimeters
1 yard (yd) = 91.44 centimeters
            = 0.9144 meter
1 mile = 1.6093 kilometers

Area

1 square centimeter (sq cm) = 0.155 square
                              inch (sq in.)
1 square meter (m²) = 1550.0 square inches
                    = 10.764 square feet
                      (sq ft)
                    = 1.196 square yards
                      (sq yd)
1 square inch = 6.4516 square centimeters
1 square yard = 0.8361 square meter

Volume

1 cubic centimeter (cc) = 0.0610 cubic inches
                          (cu in.)
1 cubic meter (m³) = 35.3145 cubic feet (cu ft)
                   = 1.3079 cubic yards (cu yd)
1 cubic inch = 16.3872 cubic centimeters
1 cubic yard = 0.7646 cubic meter

Capacity

1 milliliter (ml) = 0.2705 fluid dram (fl dr)
                  = 0.0338 fluid ounce (fl oz)
1 liter = 33.8148 fluid ounces
        = 2.1134 pints (pt)
        = 1.0567 quarts (qt)
        = 0.2642 gallon (gal)
1 fluid dram = 3.697 milliliters
1 fluid ounce = 29.573 milliliters
1 quart = 946.332 milliliters
1 gallon = 3.785 liters
1 cubic inch (cu in.) = 16.387 milliliters
1 cubic foot (cu ft) = 28.316 liters

Weight

1 gram (g) = 15.432 grains (gr)
  = 0.03527 avoirdupois ounce (avdp oz)
  = 0.03215 apothecaries' or troy ounce

1 kilogram (kg) = 35.274 avoirdupois ounces
  = 32.151 apothecaries' or troy ounces
  = 2.2046 avoirdupois pounds (avdp lb)
  = 2.6792 apothecaries' or troy pounds

1 grain = 64.7989 milligrams (mg)
1 avoirdupois ounce = 28.3495 grams
1 apothecaries' or troy ounce = 31.1035 grams
1 avoirdupois pound = 453.5924 grams
1 apothecaries' or troy pound = 373.2418 grams

Temperature

$°F = °C \times 9/5 + 32$
$°C = °F - 32 \times 5/9$

Kitchen Measure and Other Approximation

1 small test tube ≅ 30 milliliters (ml)
1 large test tube ≅ 70 milliliters
1 tumbler or 1 cup = 16 tablespoons (tbsp)
  = 8 fluid ounces (fl oz)
  = 1/2 pint (pt)
  ≅ 240 milliliters

1 teacup ≅ 4 fluid ounces = 120 milliliters
1 tablespoon ≅ 1/2 fluid ounce ≅ 16 milliliters
1 teaspoon (tsp) ≅ 4 milliliters
  ≅ 60 drops (depending on bore of medicine dropper)
1 milliliter ≅ 25 drops (depending on size of bore)

Using Coins as Substitutes for Weights

Coin	Approximate Weight (grams)
Dime	2.5
Penny	3.25
Nickel	5.0
Quarter	6.5
Half dollar	13.0
Silver dollar	26.0

(A dime is about 1 mm thick.)

Appendix VI

Suppliers of Biological Materials

Carolina Biological Supply Company is a large distributor of a variety of biological materials, including live cultures, preserved specimens, audiovisual materials, lab equipment, and many other items. They have two addresses:

Main office and laboratories
Burlington, North Carolina 27215
Phone: 919 584-0381

Powell Laboratories Division
Gladstone, Oregon 97027
Phone: 503 656-1641

Living insects and their diets are available from a number of sources. Here are a few from a listing in *Bugs: How to Raise Insects for Fun and Profit*, by Daniel and Connie Mayer, 702 S. Michigan, South Bend, Indiana 46618:

California Green Lacewings
P.O. Box 2495
Merced, CA 95340

College Biological Supply Co.
P.O. Box 25017
Northgate Station
Seattle, WA 98125

General Biological, Inc.
8200 S. Hoyne Avenue
Chicago, IL 60620

Rincon-Vitova Insectaries
P.O. Box 95
Oakview, CA 93022

Bio-Serv, Inc., P.O. Box 100B, Frenchtown, NJ 08825 is a supplier of diets for rearing insects. They supply complete diets, stock diet ingredients, vitamin mixes, salt mixes, and other equipment used in rearing insects.

Index

Page numbers in italics refer to text illustrations

t